Astrophysics: A Formula Handbook

N.B. Singh

DEDICATION

To Nature,

I dedicate this book to you, the source of all life. You are my inspiration, my teacher, and my friend.

Thank you for teaching me about the beauty of the world around me. Thank you for showing me the power of the natural world. Thank you for giving me a sense of peace and tranquillity.

I promise to do my part to protect you and your many wonders. I will teach my children about the importance of conservation and sustainability. I will work to make the world a better place for all living things.

Thank you for everything, Nature.

With love,

N.B Singh

Contents

13 Astrophysics and Society   **133**

PREFACE

Welcome to "Astrophysics: A Formula Handbook"! This handbook is designed to serve as a comprehensive reference for students, researchers, and enthusiasts delving into the fascinating realm of astrophysics.

Purpose of the Handbook

Astrophysics, a branch of astronomy, explores the celestial phenomena that shape our universe. This handbook aims to provide a concise compilation of essential formulas, equations, and concepts relevant to various aspects of astrophysics. Whether you are a student seeking a quick reference for your coursework or a researcher in need of fundamental equations, this handbook is crafted to be your go-to guide.

Organization of the Handbook

The handbook is organized into thematic chapters, each focusing on a specific aspect of astrophysics. From celestial mechanics and stellar astrophysics to cosmology and high-energy astrophysics, you will find a wealth of formulas presented in a clear and accessible manner. Mathematical notations are accompanied by brief explanations to aid understanding.

Who Can Benefit

This handbook is intended for a diverse audience. Students pursuing degrees in astronomy, astrophysics, physics, or related fields will find it a valuable companion throughout their academic journey. Researchers and professionals will appreciate the quick access to formulas needed for their work. Even enthusiasts with a passion for the cosmos will discover a wealth of information to deepen their understanding.

Using This Handbook

For optimal utility, each formula is presented with clarity and accompanied by relevant explanations. Whether you are reviewing fundamental concepts or exploring advanced topics, the handbook is structured to facilitate efficient learning and application.

I hope you find "Astrophysics: A Formula Handbook" to be a valuable resource in your exploration of the cosmos. May it accompany you on your journey through the wonders of astrophysics!

N.B. Singh

Chapter 1

Introduction

1.1 Overview

Astrophysics is the study of celestial phenomena, applying physics principles to understand the universe. Let's dive into the basics.

1.1.1 Gravity: The Cosmic Glue

The force of gravity, described by Newton's Law of Gravitation, governs celestial bodies. For two masses M_1 and M_2 separated by distance r:

$$F = G \cdot \frac{M_1 \cdot M_2}{r^2}$$

1.1.2 Einstein's Touch: General Relativity

Gravity, as explained by Einstein's General Theory of Relativity, is the curvature of spacetime caused by mass. The Einstein field equations succinctly express this cosmic dance:

$$G_{\mu\nu} + \Lambda g_{\mu\nu} = \frac{8\pi G}{c^4} T_{\mu\nu}$$

1.1.3 Cosmic Energies: E=mc²

Mass-energy equivalence, Einstein's iconic equation:

$$E = mc^2$$

This formula underlines the conversion of mass to energy and vice versa.

1.1.4   Stellar Light: Blackbody Radiation

Understanding starlight involves Planck's law:

$$B(\lambda, T) = \frac{8\pi hc}{\lambda^5} \cdot \frac{1}{e^{\frac{hc}{\lambda kT}} - 1}$$

Here, $B(\lambda, T)$ represents the spectral radiance of a blackbody at temperature T and wavelength λ.

1.1.5   Wavelength Matters: Doppler Shift

Observing cosmic motion through Doppler shift:

$$\frac{\Delta\lambda}{\lambda} = \frac{v}{c}$$

Here, $\Delta\lambda$ is the change in wavelength, λ is the initial wavelength, v is the velocity of the source, and c is the speed of light.

1.1.6   Matter in Space: Hydrostatic Equilibrium

Stellar interiors maintain equilibrium via the hydrostatic balance equation:

$$\frac{dP}{dr} = -\rho \cdot g$$

It relates the pressure gradient ($\frac{dP}{dr}$) to the gravitational acceleration (g) and density (ρ).

1.1.7   Quantum Universe: Schrödinger's Equation

At the quantum level, the wave function Ψ obeys Schrödinger's equation:

$$i\hbar\frac{\partial\Psi}{\partial t} = -\frac{\hbar^2}{2m}\nabla^2\Psi + V\Psi$$

This equation governs the evolution of quantum systems.

1.2 Historical Perspective

Explore the journey of astrophysics through key historical milestones:

1.2.1 Kepler's Laws:

Johannes Kepler unveiled the motion of planets. His laws express planetary orbits:

$$1. \quad T^2 \propto a^3$$

$$2. \quad \frac{dA}{dt} = \text{constant}$$

$$3. \quad \frac{a_1}{a_2} = \frac{T_1}{T_2}$$

T is the orbital period, a is the semi-major axis, A is the area swept, and the subscripts represent different planets.

1.2.2 Newton's Gravitation:

Isaac Newton's law of gravitation, a breakthrough in understanding celestial bodies:

$$F = G \cdot \frac{M_1 \cdot M_2}{r^2}$$

F is the gravitational force, G is the gravitational constant, M_1 and M_2 are masses, and r is the separation.

1.2.3 Einstein's Leap:

Einstein's theory of General Relativity redefined gravity:

$$G_{\mu\nu} + \Lambda g_{\mu\nu} = \frac{8\pi G}{c^4} T_{\mu\nu}$$

This equation describes the curvature of spacetime due to mass and energy.

1.2.4 Cosmic Expansion:

Edwin Hubble's discovery of the expanding universe:

$$v = H_0 \cdot d$$

The velocity (v) of galaxies is proportional to their distance (d), with H_0 as the Hubble constant.

1.2.5 Quantum Touch:

Quantum mechanics meets astrophysics through Max Planck's constant:

$$E = h \cdot f$$

Energy (E) is quantized in discrete packets, where h is Planck's constant and f is the frequency.

1.2.6 Nuclear Fusion:

Harnessing Einstein's equation for stellar energy production:

$$E = mc^2$$

Mass-energy equivalence powers stars, releasing energy in nuclear fusion.

1.2.7 Cosmic Microwave Background:

Arno Penzias and Robert Wilson's accidental discovery:

$$T_{\text{CMB}} = 2.725 \, \text{K}$$

The cosmic microwave background temperature, a remnant of the Big Bang.

1.3 Scope and Importance

Discover the vast scope and significance of astrophysics with practical insights:

1.3.1 Cosmic Scale:

Astrophysics explores the universe's immense scale, from galaxies to cosmic filaments:

$$V = \frac{4}{3}\pi R^3$$

The volume (V) of a sphere with radius (R).

1.3.2   Dark Matter Mystery:

Addressing the cosmic mystery of dark matter, crucial for understanding galactic dynamics:

$$F = m \cdot a$$

Newton's second law relates force (F), mass (m), and acceleration (a).

1.3.3   Energy in Stars:

Unveiling the energy generation within stars through nuclear fusion:

$$L = 4\pi R^2 \sigma T^4$$

The luminosity (L) of a star depends on its radius (R) and temperature (T).

1.3.4   Exoplanets:

Exploring the potential for life beyond our solar system:

$$P^2 = \frac{4\pi^2}{G(M_1 + M_2)} a^3$$

Kepler's third law relates the orbital period (P) to the semi-major axis (a) of a binary star system.

1.3.5   Big Bang Nucleosynthesis:

Understanding the origin of light elements in the early universe:

$$n \leftrightarrow p + e^- + \bar{\nu}_e$$

A neutron transforms into a proton, electron, and antineutrino, and vice versa.

1.3.6   Gravitational Waves:

Detecting ripples in spacetime, a revolutionary tool for observing cosmic events:

$$h_{\mu\nu} = \frac{2G}{c^4} \frac{\partial^2 I_{\mu\nu}}{\partial t^2}$$

The gravitational wave strain $(h_{\mu\nu})$ is related to the second derivative of the quadrupole moment.

1.3.7   Technological Spin-Offs:

Highlighting the impact of astrophysics on technology:

$$V = IR$$

Ohm's law relates voltage (V), current (I), and resistance (R).

1.4   Fundamental Concepts

Explore foundational concepts in astrophysics with simplicity:

1.4.1   Light and Colors:

Understanding the spectrum of light using wavelength (λ) and frequency (f):

$$c = \lambda \cdot f$$

The speed of light (c) equals wavelength multiplied by frequency.

1.4.2   Redshift and Blueshift:

Observing the motion of celestial objects through the Doppler effect:

$$\frac{\Delta\lambda}{\lambda} = \frac{v}{c}$$

Redshift occurs as objects move away ($v > 0$), and blueshift when approaching ($v < 0$).

1.4.3   Stellar Classification:

Categorizing stars based on temperature and spectral features:

$$T_{\text{eff}} \approx \frac{b}{\lambda_{\text{max}}}$$

The effective temperature (T_{eff}) is inversely proportional to the wavelength of maximum intensity.

1.4.4 Hertzsprung-Russell Diagram:

Mapping stars by luminosity and temperature:

$$L = 4\pi R^2 \sigma T^4$$

Luminosity (L) depends on stellar radius (R) and temperature (T).

1.4.5 Eclipsing Binaries:

Studying binary star systems through periodic light variations:

$$\frac{L_1 + L_2}{L_{\text{total}}} = \frac{1 + \cos(\pi t/P)}{2}$$

Light curve fraction for eclipsing binaries with period (P).

1.4.6 Black Holes:

Describing the boundary of a black hole, the event horizon:

$$r_s = \frac{2GM}{c^2}$$

The Schwarzschild radius (r_s) depends on the mass (M) of the black hole.

1.4.7 Dark Energy:

Quantifying the mysterious force driving the accelerated expansion of the universe:

$$p = -\rho c^2$$

Dark energy pressure (p) relates to density (ρ) and the speed of light (c).

1.5 Key Terms

Uncover essential terms in astrophysics with simplicity:

1.5.1   Magnitude and Luminosity:

Relating apparent magnitude (m) to luminosity (L) and distance (d):

$$m - M = 5 \log_{10}\left(\frac{d}{10}\right)$$

The distance modulus equation helps estimate astronomical distances.

1.5.2   Escape Velocity:

Calculating the velocity required to escape a celestial body:

$$v_{\text{escape}} = \sqrt{\frac{2GM}{R}}$$

Escape velocity depends on the gravitational constant (G), mass (M), and radius (R).

1.5.3   Red Giant Phase:

Understanding the expansion of stars in their later stages:

$$L = 4\pi R^2 \sigma T^4$$

Luminosity (L) increases during the red giant phase due to expansion.

1.5.4   Quasar Emission Lines:

Identifying spectral lines from distant, active galactic nuclei:

$$z = \frac{\lambda_{\text{observed}} - \lambda_{\text{rest}}}{\lambda_{\text{rest}}}$$

Redshift (z) in quasar spectra helps measure cosmic distances.

1.5.5   Neutron Star Mass Limit:

Defining the upper mass limit for stable neutron stars:

$$M_{\text{max}} \approx 3M_{\odot}$$

Neutron stars exceeding this limit collapse into black holes.

1.5.6 Dark Matter Detection:

Searching for dark matter through weakly interacting massive particles (WIMPs):

$$E = m_{\text{WIMP}} \cdot c^2$$

Energy (E) is related to the WIMP mass (m_{WIMP}) through mass-energy equivalence.

1.5.7 Cosmic Microwave Background (CMB):

Exploring the afterglow of the Big Bang:

$$T_{\text{CMB}} = 2.725 \, \text{K}$$

The temperature of the cosmic microwave background provides insights into the early universe.

1.6 Notable Discoveries

Uncover key discoveries in astrophysics with simplicity:

1.6.1 Hubble's Law:

Expressing the expansion of the universe discovered by Edwin Hubble:

$$v = H_0 \cdot d$$

The velocity (v) of galaxies is proportional to their distance (d), with H_0 as the Hubble constant.

1.6.2 Black Hole Thermodynamics:

Revealing the thermodynamic properties of black holes:

$$\delta M = \frac{\kappa}{8\pi G} \delta A + \Omega \delta J + \Phi \delta Q$$

The first law of black hole thermodynamics involves mass (M), surface gravity (κ), area (A), angular velocity (Ω), electric potential (Φ), and heat transfer (Q).

1.6.3   Discovery of Exoplanets:

Identifying planets beyond our solar system:

$$P^2 = \frac{4\pi^2}{G(M_1 + M_2)}a^3$$

Kepler's third law relates the orbital period (P) to the semi-major axis (a) of a binary star system.

1.6.4   Detection of Gravitational Waves:

Observing ripples in spacetime:

$$h_{\mu\nu} = \frac{2G}{c^4}\frac{\partial^2 I_{\mu\nu}}{\partial t^2}$$

The gravitational wave strain ($h_{\mu\nu}$) is related to the second derivative of the quadrupole moment.

1.6.5   Discovery of Dark Energy:

Detecting the accelerated expansion of the universe:

$$p = -\rho c^2$$

Dark energy pressure (p) relates to density (ρ) and the speed of light (c).

1.6.6   Stellar Nucleosynthesis:

Unraveling the origin of elements in stars:

$$E = mc^2$$

Einstein's mass-energy equivalence governs nuclear reactions in stars.

1.6.7   Detection of Cosmic Microwave Background (CMB):

Observing the radiation from the early universe:

$$T_{\text{CMB}} = 2.725\,\text{K}$$

The temperature of the cosmic microwave background provides insights into the early universe.

1.7 Current Challenges

Explore the forefront challenges in astrophysics with simplicity:

1.7.1 Dark Matter Understanding:

Unlocking the mystery of dark matter, comprising most of the universe:

$$\frac{dN}{dt} = -\sigma n v$$

The dark matter annihilation rate depends on the cross-section (σ), number density (n), and velocity (v).

1.7.2 Gravitational Wave Detection:

Detecting elusive gravitational waves to explore cosmic events:

$$h_{\mu\nu} = \frac{2G}{c^4} \frac{\partial^2 I_{\mu\nu}}{\partial t^2}$$

The strain ($h_{\mu\nu}$) is related to the second derivative of the quadrupole moment.

1.7.3 Understanding Dark Energy:

Unraveling the nature of dark energy driving the accelerated expansion:

$$p = -\rho c^2$$

Dark energy pressure (p) relates to density (ρ) and the speed of light (c).

1.7.4 Quantum Gravity Unification:

Uniting quantum mechanics and general relativity for a complete theory:

$$\text{Unified Field Equation: } \mathcal{L} = \mathcal{L}_{\text{gravity}} + \mathcal{L}_{\text{matter}}$$

The unified field equation includes the gravitational and matter Lagrangians.

1.7.5   Exoplanet Atmosphere Study:

Investigating exoplanet atmospheres for signs of habitability:

$$P^2 = \frac{4\pi^2}{G(M_1 + M_2)}a^3$$

Kepler's third law relates the orbital period (P) to the semi-major axis (a) of a binary star system.

1.7.6   Multi-Messenger Astronomy:

Coordinating observations across different cosmic messengers:

$$E = mc^2$$

Mass-energy equivalence relates energy (E) to mass (m) and the speed of light (c).

1.7.7   Cosmic Inflation:

Understanding the early rapid expansion of the universe:

$$H(t) = \frac{\dot{a}}{a}$$

The Hubble parameter (H) represents the rate of cosmic expansion.

Chapter 2

Celestial Mechanics

2.1 Kepler's Laws

Explore the simplicity of Kepler's Laws with practical insights:

2.1.1 Law of Orbits:

Planets move in elliptical orbits with the Sun at one focus:

$$r(\theta) = \frac{a(1 - e^2)}{1 + e\cos(\theta)}$$

The polar coordinate equation for an ellipse, where a is the semi-major axis and e is the eccentricity.

2.1.2 Law of Areas:

The line segment between a planet and the Sun sweeps equal areas during equal intervals of time:

$$\frac{dA}{dt} = \frac{1}{2}r^2\frac{d\theta}{dt}$$

The rate at which area (A) is swept equals half the product of radius (r) squared and angular velocity ($\frac{d\theta}{dt}$).

2.1.3   Law of Periods:

The square of the orbital period of a planet is proportional to the cube of its semi-major axis:

$$T^2 \propto a^3$$

The proportionality constant depends on the mass of the central body.

2.1.4   Escape Velocity:

The minimum velocity required for an object to escape the gravitational influence:

$$v_{\text{escape}} = \sqrt{\frac{2GM}{R}}$$

Escape velocity depends on the gravitational constant (G), mass (M), and radius (R).

2.1.5   Two-Body Problem:

Describing the motion of two masses under mutual gravitational attraction:

$$F = G \cdot \frac{M_1 \cdot M_2}{r^2}$$

Newton's law of gravitation relates force (F) to the masses (M_1, M_2), and distance (r) between them.

2.1.6   Three-Body Problem:

Extending gravitational interactions to systems with three masses:

$$m_1 \frac{d^2 \mathbf{r}_1}{dt^2} = \frac{G m_1 m_2 (\mathbf{r}_2 - \mathbf{r}_1)}{|\mathbf{r}_2 - \mathbf{r}_1|^3}$$

Equation of motion for mass m_1 in a three-body system, where $\mathbf{r}_1$ and $\mathbf{r}_2$ are the position vectors.

2.2   Newton's Law of Gravitation

Uncover the essence of Newton's Law of Gravitation with simplicity:

2.2.1 Gravitational Force:

Objects attract each other with a force directly proportional to the product of their masses and inversely proportional to the square of the distance between their centers:

$$F = G \cdot \frac{m_1 \cdot m_2}{r^2}$$

The gravitational force (F) is determined by the gravitational constant (G), masses $(m_1$ and $m_2)$, and the distance (r) between them.

2.2.2 Orbital Dynamics:

Understanding the motion of planets in an orbit under gravitational influence:

$$\frac{d^2\mathbf{r}}{dt^2} = -\frac{GM}{r^2}\hat{\mathbf{r}}$$

The acceleration of an object in orbit is directed towards the central mass (M), with $\hat{\mathbf{r}}$ as the unit vector pointing towards the center.

2.2.3 Escape Velocity:

Determining the minimum velocity required to escape a celestial body:

$$v_{\text{escape}} = \sqrt{\frac{2GM}{R}}$$

Escape velocity depends on the gravitational constant (G), mass (M), and the radius (R) of the celestial body.

2.2.4 Gravitational Potential Energy:

Expressing the potential energy of an object in a gravitational field:

$$U = -\frac{Gm_1m_2}{r}$$

The gravitational potential energy (U) is negative and depends on the masses $(m_1$ and $m_2)$ and their separation (r).

2.2.5   Two-Body Problem:

Describing the motion of two masses under mutual gravitational attraction:

$$F = G \cdot \frac{M_1 \cdot M_2}{r^2}$$

Newton's law of gravitation relates force (F) to the masses (M_1, M_2), and distance (r) between them.

2.2.6   Perturbation Theory:

Studying the effects of additional forces on the motion of celestial bodies:

$$m\frac{d^2\mathbf{r}}{dt^2} = -\frac{GM_1}{r^2}\hat{\mathbf{r}} + \mathbf{F}_{\text{perturbation}}$$

The equation of motion accounting for both gravitational and perturbing forces.

2.3   Orbital Dynamics

Explore the essence of Orbital Dynamics with simplicity:

2.3.1   Kepler's Third Law:

The square of the orbital period (T) is proportional to the cube of the semi-major axis (a):

$$T^2 \propto a^3$$

This relationship holds for any object orbiting a central mass.

2.3.2   Centripetal Force:

Objects in orbit experience a centripetal force keeping them in a circular path:

$$F_{\text{centripetal}} = \frac{m \cdot v^2}{r}$$

The centripetal force depends on mass (m), orbital velocity (v), and orbital radius (r).

2.3.3 Acceleration in Orbit:

The acceleration of an object in orbit points towards the central mass:

$$\frac{d^2\mathbf{r}}{dt^2} = -\frac{GM}{r^2}\hat{\mathbf{r}}$$

This acceleration ensures orbital motion and depends on the gravitational constant (G) and central mass (M).

2.3.4 Escape Velocity:

The minimum velocity required for an object to escape gravitational influence:

$$v_{\text{escape}} = \sqrt{\frac{2GM}{r}}$$

Escape velocity depends on the gravitational constant (G), central mass (M), and orbital radius (r).

2.3.5 Specific Orbital Energy:

The energy associated with an object's orbit:

$$E = \frac{1}{2}mv^2 - \frac{GMm}{r}$$

This specific orbital energy combines kinetic and potential energy terms.

2.3.6 Two-Body Problem:

Describing the motion of two masses under mutual gravitational attraction:

$$F = G \cdot \frac{M_1 \cdot M_2}{r^2}$$

Newton's law of gravitation relates force (F) to the masses (M_1, M_2), and distance (r) between them.

2.3.7   Angular Momentum in Orbit:

Angular momentum (L) is conserved for an object in orbit:

$$L = mvr$$

Angular momentum depends on mass (m), velocity (v), and orbital radius (r).

2.4   Escape Velocity

Unlock the concept of escape velocity with simplicity:

2.4.1   Definition:

Escape velocity is the minimum speed an object needs to break free from the gravitational influence of a celestial body:

$$v_{escape} = \sqrt{\frac{2GM}{R}}$$

Escape velocity (v_{escape}) depends on the gravitational constant (G), mass of the celestial body (M), and the distance from its center to the object's surface (R).

2.4.2   Practical Example:

For Earth, escape velocity is approximately 11.2 km/s. To escape Earth's gravity, a rocket or object must reach this speed.

2.4.3   Calculation:

Calculate escape velocity using the formula for any celestial body. For Earth, it simplifies to:

$$v_{escape} \approx 11.2 \, \text{km/s}$$

2.4.4   Energy Consideration:

Escape velocity arises from the balance between kinetic and gravitational potential energy:

$$\frac{1}{2}mv_{\text{escape}}^2 = \frac{GMm}{R}$$

Objects reaching escape velocity convert their gravitational potential energy into kinetic energy.

2.4.5 Applications:

Understanding escape velocity is crucial for spacecraft, ensuring they achieve the required speed for orbital insertion or interplanetary travel.

2.4.6 Escape Trajectory:

The trajectory of an object leaving a celestial body is determined by the escape velocity and initial conditions. It follows a hyperbolic path.

2.4.7 Real-world Impact:

Escape velocity is a key parameter in space missions, influencing rocket design, fuel requirements, and mission planning.

2.5 Two-Body Problem

Unveil the simplicity of the Two-Body Problem with practical insights:

2.5.1 Newton's Law of Gravitation:

Objects attract each other with a force directly proportional to the product of their masses and inversely proportional to the square of the distance between their centers:

$$F = G \cdot \frac{M_1 \cdot M_2}{r^2}$$

The gravitational force (F) depends on the gravitational constant (G), masses (M_1 and M_2), and distance (r) between them.

2.5.2   Equation of Motion:

Describing the motion of two masses under mutual gravitational attraction:

$$m\frac{d^2\mathbf{r}}{dt^2} = -\frac{GM_1}{r^2}\hat{\mathbf{r}}$$

The equation of motion for a smaller mass (m) orbiting a larger mass (M_1) at a distance (r), with $\hat{\mathbf{r}}$ as the unit vector pointing towards M_1.

2.5.3   Angular Momentum:

Angular momentum is conserved in the Two-Body Problem:

$$mrv = \text{constant}$$

The product of mass (m), radius (r), and velocity (v) remains constant throughout the motion.

2.5.4   Energy Consideration:

Total mechanical energy of the system is conserved:

$$E = \frac{1}{2}mv^2 - \frac{GM_1 m}{r}$$

The sum of kinetic and potential energy remains constant.

2.5.5   Orbital Parameters:

Orbital characteristics are determined by energy and angular momentum:

$$a = \frac{-GM_1 m}{2E}, \quad e = \sqrt{1 + \frac{2Ema}{G^2 M_1^2}}$$

The semi-major axis (a) and eccentricity (e) define the shape of the orbit.

2.5.6   Kepler's Laws:

The Two-Body Problem solutions adhere to Kepler's Laws:

1. Orbits are elliptical.

2. Equal areas are swept in equal times.

3. The square of the orbital period is proportional to the cube of the semi-major axis.

2.6 Three-Body Problem

Discover the essentials of the Three-Body Problem with simplicity:

2.6.1 Gravitational Forces:

The motion of three masses under mutual gravitational attraction is governed by Newton's law:

$$m_1 \frac{d^2\mathbf{r}_1}{dt^2} = \frac{Gm_1m_2(\mathbf{r}_2 - \mathbf{r}_1)}{|\mathbf{r}_2 - \mathbf{r}_1|^3}$$

Similar equations describe the motion of m_2 and m_3.

2.6.2 Equations of Motion:

The complete system of equations for three masses m_1, m_2, and m_3:

$$m_1 \frac{d^2\mathbf{r}_1}{dt^2} = \frac{Gm_1m_2(\mathbf{r}_2 - \mathbf{r}_1)}{|\mathbf{r}_2 - \mathbf{r}_1|^3} + \frac{Gm_1m_3(\mathbf{r}_3 - \mathbf{r}_1)}{|\mathbf{r}_3 - \mathbf{r}_1|^3}$$

$$m_2 \frac{d^2\mathbf{r}_2}{dt^2} = \frac{Gm_2m_1(\mathbf{r}_1 - \mathbf{r}_2)}{|\mathbf{r}_1 - \mathbf{r}_2|^3} + \frac{Gm_2m_3(\mathbf{r}_3 - \mathbf{r}_2)}{|\mathbf{r}_3 - \mathbf{r}_2|^3}$$

$$m_3 \frac{d^2\mathbf{r}_3}{dt^2} = \frac{Gm_3m_1(\mathbf{r}_1 - \mathbf{r}_3)}{|\mathbf{r}_1 - \mathbf{r}_3|^3} + \frac{Gm_3m_2(\mathbf{r}_2 - \mathbf{r}_3)}{|\mathbf{r}_2 - \mathbf{r}_3|^3}$$

These equations describe the accelerations of the three masses due to gravitational interactions.

2.6.3 Chaos and Predictability:

The Three-Body Problem exhibits chaotic behavior, making long-term predictions challenging.

2.6.4 Lagrange Points:

Special points in a three-body system where gravitational forces balance orbital motion:

$$\mathbf{r}_L = \mathbf{r}_1 + \frac{m_2}{m_1 + m_2}(\mathbf{r}_2 - \mathbf{r}_1)$$

These points offer stable locations for spacecraft.

2.6.5   Real-world Applications:

Understanding the Three-Body Problem is crucial for celestial mechanics, satellite dynamics, and mission planning.

2.7   Perturbation Theory

Explore Perturbation Theory with simplicity:

2.7.1   Definition:

Perturbation Theory is a mathematical approach to study the effects of additional forces on the motion of celestial bodies.

2.7.2   Equation of Motion:

The general equation of motion for a perturbed two-body problem:

$$m\frac{d^2\mathbf{r}}{dt^2} = -\frac{GM_1}{r^2}\hat{\mathbf{r}} + \mathbf{F}_{\text{perturbation}}$$

Here, $\mathbf{F}_{\text{perturbation}}$ represents additional forces beyond the gravitational force.

2.7.3   Perturbing Forces:

Common perturbing forces include the gravitational influence of other celestial bodies, atmospheric drag, or radiation pressure.

2.7.4   Small Parameter:

Perturbation Theory relies on a dimensionless small parameter, often denoted as ϵ, representing the ratio of perturbing forces to gravitational forces.

2.7.5   Perturbation Solutions:

Expressing the solution as a power series in terms of the small parameter:

$$\mathbf{r} = \mathbf{r}_0 + \epsilon\mathbf{r}_1 + \epsilon^2\mathbf{r}_2 + \dots$$

The solution is expanded in a series, where each term corresponds to a higher-order correction.

2.7.6 Laplace-Runge-Lenz Vector:

In celestial mechanics, the Laplace-Runge-Lenz vector is often used for perturbation calculations:

$$\mathbf{A} = \mathbf{p} \times \mathbf{L} - \frac{GMm}{r}\hat{\mathbf{r}}$$

Here, $\mathbf{p}$ is the linear momentum, $\mathbf{L}$ is the angular momentum, G is the gravitational constant, M is the central mass, m is the orbiting mass, and r is the distance between them.

2.7.7 Applications:

Perturbation Theory is essential for accurate predictions of celestial body trajectories, satellite orbits, and long-term behavior.

2.7.8 Real-world Impact:

Understanding perturbations is crucial for space missions, satellite navigation, and maintaining the stability of orbits.

Chapter 3

Stellar Structure

3.1 Hydrostatic Equilibrium

Grasp Hydrostatic Equilibrium with simplicity:

3.1.1 Definition:

Hydrostatic Equilibrium is the balance between the gravitational force pulling matter inward and the pressure force pushing outward in a star.

3.1.2 Equation of Hydrostatic Equilibrium:

$$\frac{dP}{dr} = -\frac{GM\rho}{r^2}$$

The rate of change of pressure (P) with respect to radius (r) equals the negative gravitational force per unit volume.

3.1.3 Pressure Force:

The pressure force supporting the star is due to thermal and radiation pressure generated by nuclear fusion in the core:

$$P = P_{\text{gas}} + P_{\text{rad}}$$

The total pressure (P) consists of gas pressure (P_{gas}) and radiation pressure (P_{rad}).

3.1.4   Internal Energy Source:

Nuclear fusion in the star's core provides the internal energy source:

$$L = 4\pi r^2 \sigma T^4$$

Luminosity (L) depends on the stellar radius (r) and temperature (T).

3.1.5   Mass Continuity Equation:

Relating the change in mass with radius:

$$\frac{dm}{dr} = 4\pi r^2 \rho$$

The mass continuity equation ensures conservation of mass in the stellar structure.

3.1.6   Ideal Gas Law:

Relating pressure, density, and temperature for the stellar gas:

$$P_{\text{gas}} = \frac{\rho k T}{\mu m_u}$$

Where k is the Boltzmann constant, μ is the mean molecular weight, and m_u is the atomic mass unit.

3.1.7   Energy Transport:

The three methods of energy transport in a star:

- Radiative Diffusion: Energy transported by photons.

- Convection: Energy transported by moving masses.

- Conductive Transport: Energy transferred by collisions between particles.

3.1.8   Solar Core Temperature:

The core temperature of the Sun, sustained by nuclear fusion:

$$T_{\text{core}} \approx 15 \, \text{million K}$$

The high core temperature enables nuclear fusion reactions.

3.2 Nuclear Fusion

Grasp Nuclear Fusion with simplicity:

3.2.1 Definition:

Nuclear Fusion is the process by which lightweight atomic nuclei combine to form heavier nuclei, releasing energy.

3.2.2 Proton-Proton Chain:

In the Sun's core, the primary fusion process is the Proton-Proton Chain:

$$4H \rightarrow He + 2e^+ + 2\nu_e$$

Four hydrogen nuclei (protons) combine to form one helium nucleus, positrons (e^+), and neutrinos (ν_e).

3.2.3 Energy Release:

The energy released in nuclear fusion is given by Einstein's mass-energy equivalence principle:

$$E = \Delta m \cdot c^2$$

Where Δm is the change in mass, and c is the speed of light.

3.2.4 Binding Energy:

The binding energy per nucleon is a measure of nuclear stability:

$$BE = \frac{\text{Total Binding Energy}}{\text{Number of Nucleons}}$$

Higher binding energy per nucleon indicates greater nuclear stability.

3.2.5   Solar Luminosity:

The total energy output of the Sun, mainly originating from nuclear fusion:

$$L = 4\pi r^2 \sigma T^4$$

Luminosity (L) depends on the stellar radius (r) and temperature (T).

3.2.6   CNO Cycle:

An alternative fusion process in massive stars, the Carbon-Nitrogen-Oxygen (CNO) cycle:

$$^{12}\mathrm{C} + \mathrm{H} \rightarrow {}^{13}\mathrm{N} + \gamma$$

The CNO cycle involves the conversion of carbon and hydrogen into nitrogen and gamma rays.

3.2.7   Solar Neutrinos:

Neutrinos produced in solar fusion reactions, crucial for studying the Sun's internal processes:

$$\nu_e + p \rightarrow n + e^+$$

The production of solar neutrinos through the interaction of electron neutrinos with protons.

3.2.8   Nuclear Stability:

The balance between the forces of nuclear attraction and electrostatic repulsion determines nuclear stability.

3.2.9   Energy Transport:

Energy generated by nuclear fusion is transported outward through the stellar interior via radiative diffusion and convection.

3.3   Stellar Evolution

Grasp Stellar Evolution with simplicity:

3.3.1 Hydrogen Depletion:

As a star exhausts its core hydrogen, it undergoes changes leading to the next stages of stellar evolution.

3.3.2 Helium Core Fusion:

In more massive stars, helium fusion begins in the core, forming carbon and oxygen:

$$3\,^4\text{He} \rightarrow\, ^{12}\text{C} + \gamma$$

Helium fusion releases energy, maintaining pressure and preventing gravitational collapse.

3.3.3 Red Giant Phase:

As the core contracts, the outer envelope expands, turning the star into a red giant.

3.3.4 Helium Flash:

In low-mass stars, helium fusion initiates explosively, known as a helium flash, restoring stability.

3.3.5 Planetary Nebula Formation:

During the red giant phase, the outer layers are expelled, forming a planetary nebula:

$$C + O \rightarrow CO + \gamma$$

The expelled material enriches the interstellar medium with heavy elements.

3.3.6 White Dwarf Formation:

The remaining core contracts to form a white dwarf, a dense, Earth-sized remnant composed of degenerate matter.

3.3.7 Supernova Explosion:

Massive stars undergo a supernova explosion when their cores collapse and rebound, leading to the ejection of outer layers:

$$Fe \rightarrow Ni + \gamma$$

The energy released in a supernova produces heavy elements and contributes to cosmic ray acceleration.

3.3.8   Neutron Star or Black Hole:

The core remnants of a supernova collapse further to form a neutron star or, in the case of higher mass, a black hole.

3.3.9   Stellar Remnants:

White dwarfs, neutron stars, and black holes are the final evolutionary stages, influencing the composition of the universe.

3.3.10   Stellar Lifetimes:

Stellar lifetimes vary based on mass. Massive stars have shorter lifespans due to higher energy consumption.

3.3.11   Stellar Feedback:

Stellar evolution influences the surrounding environment through processes like stellar winds, enriching the interstellar medium.

3.4   Stellar Equilibrium

Grasp Stellar Equilibrium with simplicity:

3.4.1   Pressure and Gravity Balance:

In a star, equilibrium is maintained by the balance between gravitational forces pulling inward and pressure forces pushing outward.

3.4.2 Hydrostatic Equilibrium:

Expressed by the hydrostatic equilibrium equation:

$$\frac{dP}{dr} = -\frac{GM\rho}{r^2}$$

The rate of pressure change with radius equals the negative gravitational force per unit volume.

3.4.3 Luminosity and Energy Transport:

The luminosity (L) of a star is the energy transported from the core to the surface. It is determined by nuclear fusion and expressed as:

$$L = 4\pi r^2 \sigma T^4$$

Where r is the stellar radius and T is the temperature.

3.4.4 Mass Continuity Equation:

Describing how mass changes with radius:

$$\frac{dm}{dr} = 4\pi r^2 \rho$$

The mass continuity equation ensures the conservation of mass in stellar structure.

3.4.5 Ideal Gas Law:

Relating pressure, density, and temperature in the stellar interior:

$$P = \frac{\rho k T}{\mu m_u}$$

Where P is pressure, ρ is density, k is the Boltzmann constant, μ is the mean molecular weight, and m_u is the atomic mass unit.

3.4.6 Stellar Core Temperature:

Determined by nuclear fusion reactions in the core:

$$T_{\text{core}} \approx 15\,\text{million K}$$

The high core temperature enables nuclear fusion, maintaining stellar equilibrium.

3.4.7   Energy Transport Mechanisms:

Three methods of energy transport in a star:

- Radiative Diffusion: Energy transported by photons.

- Convection: Energy transported by moving masses.

- Conductive Transport: Energy transferred by collisions between particles.

3.4.8   Stellar Stability:

Stellar equilibrium ensures stability against gravitational collapse or runaway expansion, maintaining a balanced structure.

3.5   Mass-Luminosity Relation

Grasp the Mass-Luminosity Relation with simplicity:

3.5.1   Definition:

The Mass-Luminosity Relation describes the connection between a star's mass (M) and its luminosity (L).

3.5.2   Equation:

The relation is often expressed as a power-law relationship:

$$L \propto M^{\alpha}$$

Where α is an exponent that depends on the star's evolutionary stage.

3.5.3   Main Sequence Relation:

For stars in the main sequence phase, the Mass-Luminosity Relation is approximately:

$$L \approx M^{3.5}$$

This relation indicates that more massive stars have significantly higher luminosities.

3.5.4 Stellar Evolution Influence:

As a star evolves, the Mass-Luminosity Relation changes, reflecting the different energy generation processes in the stellar core.

3.5.5 Limitations:

The Mass-Luminosity Relation is an empirical relationship and may not hold for stars in other evolutionary stages or extreme conditions.

3.5.6 Observational Confirmation:

Observational data from star clusters and binary star systems support the general trend of the Mass-Luminosity Relation.

3.5.7 Real-world Application:

Understanding the Mass-Luminosity Relation is essential for estimating the luminosities of distant stars based on their observed masses.

3.5.8 Implications:

The relation has implications for the classification and understanding of stars, helping astronomers interpret observational data.

3.5.9 Varied Exponents:

In different wavelength bands or for stars in specific phases, the exponent in the Mass-Luminosity Relation may vary, leading to nuanced relationships.

3.5.10   Research Area:

Investigating the Mass-Luminosity Relation continues to be a significant area of research in astrophysics, contributing to our understanding of stellar properties.

3.6   Stellar Atmospheres

Grasp Stellar Atmospheres with simplicity:

3.6.1   Radiative Transfer Equation:

Describes how radiation moves through a stellar atmosphere:

$$\frac{dI_\nu}{ds} = -\alpha_\nu I_\nu + \eta_\nu$$

Where I_ν is the specific intensity, s is distance, α_ν is the absorption coefficient, and η_ν is the emission coefficient.

3.6.2   Eddington Approximation:

Simplifies the radiative transfer equation for optically thick atmospheres:

$$\frac{dI_\nu}{ds} = -\alpha_\nu I_\nu + \eta_\nu \frac{2}{3}$$

This approximation is valid in regions where scattering dominates.

3.6.3   Stellar Spectra:

The observed spectrum of a star is influenced by its atmosphere, revealing absorption lines indicative of chemical elements present.

3.6.4   Hydrogen Absorption Lines:

In the visible spectrum, Balmer lines indicate the presence of hydrogen in a star's atmosphere.

3.6.5 Stellar Absorption Features:

Other elements, such as helium, calcium, and iron, contribute distinctive absorption features to a star's spectrum.

3.6.6 Stellar Colors:

The color of a star depends on its effective temperature, with hotter stars appearing blue and cooler stars appearing red.

3.6.7 Stellar Limb Darkening:

The center of a star's disk appears brighter than its limb due to the variation in optical depth across the stellar atmosphere.

3.6.8 Blackbody Radiation:

Stellar atmospheres are often approximated as blackbodies, emitting radiation according to Planck's law:

$$B_\nu(T) = \frac{2h\nu^3}{c^2} \frac{1}{e^{h\nu/kT} - 1}$$

Where T is the temperature, ν is the frequency, h is Planck's constant, and k is the Boltzmann constant.

3.6.9 Stellar Opacity:

Opacity measures the resistance of a medium to the flow of radiation. The Rosseland mean opacity is often used in stellar models.

3.6.10 Real-world Application:

Understanding stellar atmospheres is crucial for interpreting observed spectra, classifying stars, and determining their fundamental properties.

3.7   Stellar Nucleosynthesis

Grasp Stellar Nucleosynthesis with simplicity:

3.7.1   Hydrogen Fusion:

The primary process in stellar nucleosynthesis occurs in a star's core through hydrogen fusion:

$$4\,^1\text{H} \rightarrow\,^4\text{He} + 2e^+ + 2\nu_e$$

Four hydrogen nuclei combine to form helium, releasing positrons (e^+) and neutrinos (ν_e).

3.7.2   Helium Fusion:

In more massive stars, helium fusion produces heavier elements:

$$3\,^4\text{He} \rightarrow\,^{12}\text{C} + \gamma$$

Three helium nuclei combine to form carbon, releasing gamma rays.

3.7.3   Carbon Fusion:

In even more massive stars, carbon fusion leads to the formation of heavier elements:

$$^{12}\text{C} +\,^{12}\text{C} \rightarrow\,^{24}\text{Mg} + \gamma$$

Two carbon nuclei combine to form magnesium, releasing gamma rays.

3.7.4   Supernova Nucleosynthesis:

During a supernova explosion, rapid nucleosynthesis occurs, creating elements beyond iron:

$$\text{Fe} \rightarrow \text{Ni} + \gamma$$

Iron fusion is endothermic, absorbing energy.

3.7.5   Alpha Process:

In the cores of massive stars, the alpha process produces alpha particles (^{4}He) from the fusion of lighter elements.

3.7.6 S-Process and R-Process:

Slow and rapid neutron capture processes in evolved stars contribute to the synthesis of heavy elements beyond iron.

3.7.7 Real-world Impact:

Stellar nucleosynthesis enriches the interstellar medium with a variety of elements, influencing the formation of planets and life.

3.7.8 Isotope Abundances:

The observed abundances of isotopes in the universe reflect the balance between nucleosynthesis and subsequent stellar processes.

3.7.9 Nuclear Binding Energy:

The energy released or absorbed during nucleosynthesis is determined by the nuclear binding energy of the involved elements.

3.7.10 Nuclear Reaction Rates:

The rates of nuclear reactions depend on temperature and particle energies within stellar cores.

Chapter 4

Galactic Astronomy

4.1 Galactic Structure

Grasp Galactic Structure with simplicity:

4.1.1 Milky Way Components:

The Milky Way consists of a central bulge, a disk with spiral arms, and a surrounding halo.

4.1.2 Stellar Orbits:

Stars in the galactic disk orbit the center in nearly circular paths, while those in the halo follow more elliptical orbits.

4.1.3 Spiral Arm Formation:

Density waves in the galactic disk trigger the formation of spiral arms, where new stars are born.

4.1.4 Galactic Rotation Curve:

The rotational velocity of stars in the Milky Way doesn't follow expectations based on visible matter, indicating the presence of dark matter.

4.1.5   Mass-Luminosity Ratio:

The mass-to-luminosity ratio in galaxies helps estimate the amount of dark matter present.

4.1.6   Galactic Coordinates:

Celestial objects are located using galactic coordinates, specifying positions in the Milky Way.

4.1.7   Interstellar Medium (ISM):

The ISM contains gas and dust between stars, influencing star formation and providing raw materials.

4.1.8   Hydrogen Emission Lines:

Mapping the distribution of neutral hydrogen using 21-cm emission lines reveals the spiral structure of the Milky Way.

4.1.9   Star Formation Rate:

The rate at which new stars form in a galaxy depends on the density of interstellar gas.

4.1.10   Chemical Composition:

Stars in the Milky Way have varying chemical compositions, reflecting the history of nucleosynthesis.

4.1.11   Stellar Populations:

Stars are classified into populations based on age, metallicity, and location within the galaxy.

4.1.12   Galactic Center:

The Milky Way's central region contains a supermassive black hole, Sagittarius A*.

4.1.13   Real-world Applications:

Understanding galactic structure helps interpret observations and develop models for other spiral galaxies.

4.1.14 Galactic Evolution:

The study of galactic structure provides insights into the evolution and dynamics of galaxies over cosmic time.

4.2 Spiral Arms

Grasp Spiral Arms with simplicity:

4.2.1 Formation Mechanism:

Density waves in the galactic disk trigger the formation of spiral arms.

4.2.2 Mathematical Representation:

The density wave theory can be mathematically expressed as:

$$m\frac{\partial^2 \Phi}{\partial t^2} + 3m\Omega\frac{\partial \Phi}{\partial t} + \frac{\partial \Phi}{\partial \theta} = -\frac{\partial \Phi}{\partial r}$$

Where Φ is the gravitational potential, m is the number of arms, Ω is the angular frequency, and (r, θ) are polar coordinates.

4.2.3 Star Formation in Arms:

Spiral arms are regions of enhanced star formation due to compression of interstellar gas.

4.2.4 Ongoing Research:

Understanding the dynamics of spiral arms is an active area of research in galactic astronomy.

4.2.5 Observable Features:

Observations reveal young, hot stars, gas, and dust concentrated in spiral arms.

4.2.6 Molecular Clouds:

Molecular clouds within spiral arms serve as the birthplaces of new stars.

4.2.7 HII Regions:

Ionized hydrogen regions (HII regions) are common in spiral arms, indicating recent star formation.

4.2.8 Stellar Orbits:

Stars within spiral arms orbit the galactic center, influenced by the density wave.

4.2.9 Spiral Arm Patterns:

Different galaxies exhibit various patterns and numbers of spiral arms.

4.2.10 Galactic Bar Influence:

Some spiral galaxies have central bars that can affect the formation and structure of spiral arms.

4.2.11 Gravitational Resonance:

Resonance with the bar or other galactic features influences the stability and persistence of spiral arms.

4.2.12 Real-world Impact:

Understanding spiral arms contributes to our knowledge of galaxy evolution and star formation processes.

4.2.13 Photometric Studies:

Photometric studies help map the distribution of stars and star-forming regions within spiral arms.

4.3 Stellar Populations

Grasp Stellar Populations with simplicity:

4.3.1 Definition:

Stellar populations are groups of stars with similar properties, such as age and chemical composition.

4.3.2 Mathematical Expression:

The luminosity function, describing the distribution of luminosities in a stellar population, is given by:

$$\Phi(L) = \frac{dN}{dL}$$

Where $\Phi(L)$ is the luminosity function, and dN is the number of stars in a luminosity range dL.

4.3.3 Types of Stellar Populations:

Stellar populations are categorized into three main types: Populations I, II, and III.

4.3.4 Population I Stars:

These stars are relatively young, metal-rich, and commonly found in the galactic disk.

4.3.5 Population II Stars:

Older stars with lower metallicity, often found in the galactic halo and bulge.

4.3.6 Population III Stars:

Theoretical and hypothetical stars with extremely low metallicity, considered to be the first stars formed in the universe.

4.3.7 Color-Magnitude Diagrams (CMDs):

CMDs are used to classify stars based on their colors and magnitudes, providing information about stellar populations.

4.3.8 Star Clusters:

Globular and open clusters are excellent examples of stellar populations, with stars formed from the same molecular cloud.

4.3.9   Chemical Abundances:

Stellar spectra reveal the chemical composition of stars, aiding in the classification of stellar populations.

4.3.10   Kinematics:

The study of stellar motions, such as radial velocities and proper motions, helps identify different populations.

4.3.11   Stellar Ages:

Stellar populations can be aged using isochrone fitting, comparing theoretical models with observed data.

4.3.12   Real-world Applications:

Understanding stellar populations contributes to our knowledge of galaxy formation, evolution, and cosmology.

4.3.13   Galactic Archaeology:

Studying stellar populations helps astronomers unravel the history of galaxies and the universe.

4.4   Galactic Rotation

Grasp Galactic Rotation with simplicity:

4.4.1   Rotational Velocity:

The rotational velocity of stars in a galaxy can be determined using the equation:

$$v = R \cdot \Omega$$

Where v is the rotational velocity, R is the distance from the galactic center, and Ω is the angular frequency.

4.4.2 Observed Galactic Rotation:

Observations of galactic rotation curves show that rotational velocities remain relatively constant with increasing distance from the center.

4.4.3 Dark Matter Influence:

The flat rotation curves indicate the presence of dark matter in galaxies, contributing to the total mass.

4.4.4 Modified Newtonian Dynamics (MOND):

MOND is an alternative theory proposed to explain galactic rotation curves without invoking dark matter.

4.4.5 Keplerian Motion:

In the inner regions of a galaxy, where the gravitational field dominates, stars follow Keplerian motion.

4.4.6 Differential Galactic Rotation:

Stars at different distances from the galactic center have varying orbital periods, leading to differential rotation.

4.4.7 Tidal Forces:

Tidal forces between stars can affect galactic rotation, especially in interacting galaxies or galactic clusters.

4.4.8 Galactic Bar Influence:

Central bars in galaxies can alter the rotation pattern, influencing the overall galactic dynamics.

4.4.9 Kinematic Methods:

Doppler shifts and proper motions are kinematic methods used to measure galactic rotation.

4.4.10   Real-world Impact:

Understanding galactic rotation is crucial for studying galaxy structure, dynamics, and the distribution of matter.

4.4.11   Future Observations:

Advancements in observational techniques and instruments continue to refine our understanding of galactic rotation.

4.5   Dark Matter

Grasp Dark Matter with simplicity:

4.5.1   Observational Evidence:

Galactic rotation curves and gravitational lensing provide strong evidence for the existence of dark matter.

4.5.2   Mass Discrepancy:

The observed mass in galaxies, based on luminous matter, is insufficient to explain gravitational effects, indicating the presence of dark matter.

4.5.3   Modified Gravity Theories:

Dark matter is a placeholder term for the unknown mass contributing to galactic dynamics, not explained by modified gravity theories.

4.5.4   Dark Matter Halo:

Galaxies are surrounded by extended halos of dark matter, influencing their overall structure and rotation curves.

4.5.5 Dark Matter Density Profile:

The density profile of dark matter in galaxies is often modeled using the Navarro-Frenk-White (NFW) profile:

$$\rho(r) = \frac{\rho_0}{\frac{r}{R_s}\left(1 + \frac{r}{R_s}\right)^2}$$

Where $\rho(r)$ is the dark matter density, ρ_0 is a scale density, r is the radial distance, and R_s is the scale radius.

4.5.6 WIMP Candidates:

Weakly Interacting Massive Particles (WIMPs) are popular candidates for dark matter, interacting weakly with ordinary matter.

4.5.7 Dark Matter Detection:

Efforts to detect dark matter include experiments in underground laboratories and astrophysical observations.

4.5.8 Cosmological Significance:

Dark matter plays a crucial role in the large-scale structure and evolution of the universe.

4.5.9 Cosmic Microwave Background (CMB):

The imprint of dark matter is seen in the anisotropies of the CMB, providing insights into its distribution.

4.5.10 Real-world Applications:

Understanding dark matter is vital for comprehending the cosmic web, galaxy formation, and the fate of the universe.

4.5.11   Ongoing Research:

Research continues to unravel the nature of dark matter, with experiments and observations pushing the boundaries of our understanding.

4.6   Galactic Dynamics

Grasp Galactic Dynamics with simplicity:

4.6.1   Newtonian Gravity:

The gravitational force between two masses m_1 and m_2 separated by distance r is given by Newton's law of gravitation:

$$F = G\frac{m_1 \cdot m_2}{r^2}$$

Where F is the gravitational force, G is the gravitational constant.

4.6.2   Kepler's Laws:

In a galactic system, stars follow Keplerian orbits governed by Kepler's laws of planetary motion.

4.6.3   Escape Velocity:

The escape velocity from a galactic system is given by:

$$v_{\text{esc}} = \sqrt{\frac{2GM}{r}}$$

Where v_{esc} is the escape velocity, G is the gravitational constant, M is the total mass, and r is the distance from the center.

4.6.4   Galactic Potential Energy:

The gravitational potential energy of a star in a galactic system is:

$$U = -\frac{GMm}{r}$$

Where U is the potential energy, G is the gravitational constant, M is the total mass, m is the star's mass, and r is the distance from the center.

4.6.5 Centrifugal Force:

The centrifugal force acting on a star in orbit is given by:

$$F_{\text{cent}} = \frac{mv^2}{r}$$

Where F_{cent} is the centrifugal force, m is the star's mass, v is its velocity, and r is the distance from the center.

4.6.6 Virial Theorem:

The virial theorem relates the kinetic and potential energy in a galactic system:

$$2\langle T \rangle = -\langle U \rangle$$

Where $\langle T \rangle$ is the time-averaged kinetic energy, and $\langle U \rangle$ is the time-averaged potential energy.

4.6.7 Dark Matter Contribution:

Observations of galactic dynamics suggest the presence of dark matter, influencing the overall gravitational field.

4.6.8 Real-world Impact:

Understanding galactic dynamics is crucial for unraveling the structure, evolution, and behavior of galaxies.

4.6.9 Observational Techniques:

Kinematic studies, such as spectroscopy, provide valuable insights into the motion and dynamics of stars within galaxies.

4.7 Galactic Center

Grasp the Galactic Center with simplicity:

4.7.1   Central Supermassive Black Hole:

At the heart of our Milky Way lies a supermassive black hole, Sagittarius A* (Sgr A*).

4.7.2   Black Hole Mass:

The mass of Sgr A* is estimated to be around 4×10^6 times that of our Sun.

4.7.3   Accretion Disk:

Surrounding the black hole, an accretion disk forms as material spirals in, emitting X-rays and other radiation.

4.7.4   Orbital Motion:

Stars near the galactic center exhibit rapid orbital motion, allowing astronomers to deduce the presence of the central black hole.

4.7.5   Keplerian Orbits:

The observed Keplerian orbits of stars near the galactic center confirm the presence of a massive, compact object.

4.7.6   Gravitational Redshift:

The gravitational redshift in the spectra of stars near Sgr A* provides further evidence for a massive central object.

4.7.7   Mathematical Formulas:

The gravitational force near a black hole is given by Newton's law of gravitation:

$$F = \frac{G \cdot M \cdot m}{r^2}$$

Where F is the gravitational force, G is the gravitational constant, M is the mass of the black hole, m is the mass of the star, and r is the distance between them.

4.7.8 Real-world Applications:

Understanding the galactic center enhances our knowledge of black hole dynamics and galactic evolution.

4.7.9 Infrared Observations:

Infrared observations are crucial for studying the galactic center, as dust obscures visible light.

4.7.10 Future Observations:

Advancements in observational technologies continue to unveil the mysteries of the galactic center, offering insights into the evolution of galaxies.

Chapter 5

Cosmology

5.1 Big Bang Theory

Grasp the Big Bang Theory with simplicity:

5.1.1 Definition:

The Big Bang Theory describes the origin and evolution of the universe from an extremely hot and dense state.

5.1.2 Mathematical Formulas:

Friedmann Equations:

The Friedmann equations govern the dynamics of the expanding universe:

$$H^2 = \frac{8\pi G}{3}\rho - \frac{k}{a^2}$$

$$\frac{\ddot{a}}{a} = -\frac{4\pi G}{3}(\rho + 3p)$$

Where H is the Hubble parameter, G is the gravitational constant, ρ is the energy density, k is the curvature parameter, a is the scale factor, and p is the pressure.

Hubble's Law:

Hubble's law relates the recession velocity of galaxies to their distance:

$$v = H_0 \cdot d$$

Where v is the recession velocity, H_0 is the Hubble constant, and d is the distance.

5.1.3   Observable Evidence:

Cosmic Microwave Background (CMB):

The CMB is a remnant radiation from the early universe, supporting the Big Bang Theory.

Abundance of Light Elements:

The observed abundance of light elements (primarily hydrogen and helium) aligns with predictions from Big Bang nucleosynthesis.

Redshift of Galaxies:

The redshift of galaxy spectra provides evidence for the expansion of the universe and supports the Big Bang model.

5.1.4   Inflationary Universe:

The inflationary model proposes a brief period of exponential expansion, resolving certain cosmological puzzles.

5.1.5   Cosmological Parameters:

Hubble Constant:

The current rate of expansion of the universe, denoted as H_0, is a key parameter.

Density Parameters:

Ω_m, Ω_Λ, and Ω_k represent the density contributions from matter, dark energy, and curvature, respectively.

5.1.6 Real-world Impact:

The Big Bang Theory revolutionized our understanding of the universe's history and set the foundation for modern cosmology.

5.1.7 Future Observations:

Ongoing and future observations continue to refine our understanding of the early universe and its evolution.

5.2 Cosmic Microwave Background

Grasp the Cosmic Microwave Background with simplicity:

5.2.1 Definition:

The Cosmic Microwave Background (CMB) is the faint glow of radiation filling the universe, resulting from the hot, dense state shortly after the Big Bang.

5.2.2 Blackbody Radiation:

The CMB is a nearly perfect blackbody radiation with a temperature of approximately 2.7 Kelvin.

5.2.3 Mathematical Formulas:

Planck's Law:

Planck's law describes the spectrum of blackbody radiation:

$$B(\nu, T) = \frac{8\pi h \nu^3}{c^3} \cdot \frac{1}{e^{\frac{h\nu}{kT}} - 1}$$

Where B is the spectral radiance, ν is the frequency, T is the temperature, h is Planck's constant, c is the speed of light, and k is the Boltzmann constant.

5.2.4   Observable Features:

Anisotropies:

Small temperature variations in the CMB, or anisotropies, provide crucial information about the early universe's density fluctuations.

CMB Dipole Anomaly:

The CMB exhibits a dipole anisotropy due to the Earth's motion through space.

CMB Polarization:

Polarization patterns in the CMB reveal information about the universe's geometry and gravitational waves.

5.2.5   Experimental Confirmations:

COBE Satellite:

The Cosmic Background Explorer (COBE) satellite made the first precise measurements of the CMB anisotropies.

WMAP and Planck Missions:

Successor missions like the Wilkinson Microwave Anisotropy Probe (WMAP) and Planck provided detailed maps of the CMB.

BICEP/Keck:

Experiments like BICEP/Keck aim to detect primordial gravitational waves imprinted on the CMB.

5.2.6   Cosmological Significance:

The CMB is a powerful tool for studying the composition, age, and geometry of the universe.

5.2.7   Real-world Impact:

Observations of the CMB have led to key insights, supporting the Big Bang Theory and shaping our understanding of the cosmos.

5.3   Dark Energy

Grasp Dark Energy with simplicity:

5.3.1   Definition:

Dark energy is an unknown form of energy that permeates space and is responsible for the accelerated expansion of the universe.

5.3.2   Mathematical Formulas:

Cosmological Constant:

Einstein's cosmological constant (Λ) represents a constant energy density filling space, contributing to dark energy. It is included in the Friedmann equations:

$$H^2 = \frac{8\pi G}{3}\rho + \frac{\Lambda}{3} - \frac{k}{a^2}$$

$$\frac{\ddot{a}}{a} = -\frac{4\pi G}{3}(\rho + 3p) + \frac{\Lambda}{3}$$

Where H is the Hubble parameter, G is the gravitational constant, ρ is the energy density, k is the curvature parameter, a is the scale factor, p is the pressure, and Λ is the cosmological constant.

Equation of State:

The equation of state for dark energy is often represented as $p = w\rho$, where w is the equation of state parameter.

5.3.3   Observable Effects:

Cosmic Acceleration:

Dark energy is inferred from the accelerated expansion of the universe, observed through redshift-distance measurements.

Supernova Observations:

Type Ia supernovae are used as standard candles to measure cosmic distances and infer dark energy's influence.

5.3.4   Real-world Impact:

Understanding dark energy is crucial for predicting the fate of the universe and refining cosmological models.

5.3.5   Dark Energy Models:

Quintessence:

Quintessence models propose a dynamic, evolving field as the source of dark energy.

Modified Gravity Theories:

Some theories suggest modifications to gravity on cosmological scales as an alternative to dark energy.

5.3.6   Current Challenges:

The nature of dark energy remains a significant mystery in modern cosmology, and ongoing research seeks to unravel its properties.

5.4   Dark Matter in the Universe

Grasp Dark Matter with simplicity:

5.4.1   Definition:

Dark matter is a mysterious, non-luminous substance that outweighs visible matter and influences the large-scale structure of the universe.

5.4.2   Mathematical Formulas:

Dark Matter Density:

The density of dark matter in the universe can be described by the equation:

$$\rho_{\text{DM}}(a) = \frac{\rho_{\text{DM,0}}}{a^3}$$

Where $\rho_{\mathrm{DM}}(a)$ is the dark matter density at scale factor a, and $\rho_{\mathrm{DM},0}$ is the present-day dark matter density.

Dark Matter Halo Profile:

The NFW (Navarro-Frenk-White) profile is commonly used to model the density distribution of dark matter halos:

$$\rho(r) = \frac{\rho_0}{\frac{r}{r_s}\left(1 + \frac{r}{r_s}\right)^2}$$

Where $\rho(r)$ is the dark matter density, ρ_0 is a scale density, r is the radial distance, and r_s is the scale radius.

5.4.3 Observable Effects:

Gravitational Lensing:

Dark matter's gravitational influence causes gravitational lensing, bending the path of light from background objects.

Galactic Rotation Curves:

Observations of galactic rotation curves indicate the presence of dark matter, as the rotational velocities remain constant at large radii.

Large-Scale Structure:

Dark matter is crucial for the formation of large-scale structures in the universe, such as galaxy clusters and cosmic filaments.

5.4.4 Real-world Impact:

Understanding dark matter is essential for explaining observed astronomical phenomena and the overall cosmic web structure.

5.4.5 Dark Matter Candidates:

Weakly Interacting Massive Particles (WIMPs):

WIMPs are one of the leading candidates for dark matter, interacting weakly with ordinary matter.

Axions:

Axions are hypothetical particles with low mass and are considered potential dark matter candidates.

5.4.6 Current Challenges:

Despite its significant role, the true nature of dark matter remains a major unsolved puzzle in astrophysics and particle physics.

5.5 Formation of Cosmic Structures

Grasp Cosmic Structure Formation with simplicity:

5.5.1 Definition:

Cosmic structure formation refers to the processes by which galaxies, galaxy clusters, and larger structures evolved from small density fluctuations in the early universe.

5.5.2 Mathematical Formulas:

Linear Perturbation Theory:

The growth of cosmic structures in the linear regime can be described by the linear growth factor, $D(a)$:

$$\frac{d^2 D}{da^2} + \frac{2}{a}\frac{dD}{da} - \frac{3}{2}\Omega_m H_0^2 D = 0$$

Where a is the scale factor, Ω_m is the matter density parameter, and H_0 is the Hubble constant.

Zeldovich Approximation:

The Zeldovich approximation provides a simple way to model the initial stages of structure formation, describing the displacement field $\vec{S}$:

$$\vec{S}(\vec{q}, a) = -D(a)\nabla\Phi_0(\vec{q})$$

Where $\vec{q}$ is the initial position, a is the scale factor, $D(a)$ is the growth factor, and $\Phi_0(\vec{q})$ is the gravitational potential.

5.5.3 Stages of Formation:

Linear Regime:

In the early universe, density fluctuations are small, and linear perturbation theory accurately describes the growth of structures.

Nonlinear Regime:

As structures grow, gravitational interactions lead to a transition to the nonlinear regime, forming cosmic structures like galaxies and clusters.

5.5.4 Observable Consequences:

Large-Scale Structure:

Observations of the large-scale distribution of galaxies and galaxy clusters provide insights into the cosmic web's structure.

Cosmic Microwave Background (CMB):

The CMB carries imprints of primordial density fluctuations, revealing information about the early stages of structure formation.

5.5.5 Real-world Impact:

Understanding cosmic structure formation is crucial for comprehending the universe's large-scale organization and evolution.

5.5.6 Numerical Simulations:

Cutting-edge numerical simulations, like N-body simulations, play a pivotal role in modeling and understanding the detailed processes of cosmic structure formation.

5.6   Cosmic Inflation

Grasp Cosmic Inflation with simplicity:

5.6.1   Definition:

Cosmic inflation is a hypothetical, rapid expansion of the early universe that solves several cosmological puzzles.

5.6.2   Mathematical Formulas:

Inflationary Potential:

The inflationary potential, $V(\phi)$, is a scalar field potential driving inflation. It's often represented by the slow-roll potential:

$$V(\phi) = V_0 \exp\left(-\frac{\phi}{M_p}\right)$$

Where ϕ is the inflaton field, M_p is the Planck mass, and V_0 is a constant.

Friedmann Equations during Inflation:

The Friedmann equations during inflation are modified to include the energy density of the inflaton field:

$$H^2 = \frac{1}{3M_p^2}\left(\frac{1}{2}\dot{\phi}^2 + V(\phi)\right)$$

$$\ddot{\phi} + 3H\dot{\phi} + V'(\phi) = 0$$

Where H is the Hubble parameter, $\dot{\phi}$ is the time derivative of the inflaton field, $V(\phi)$ is the inflationary potential, and $V'(\phi)$ is the derivative of V with respect to ϕ.

5.6.3   Observable Consequences:

Cosmic Microwave Background (CMB) Anisotropies:

Inflation predicts specific patterns in the CMB, explaining the observed uniformity and isotropy.

Large-Scale Structure Formation:

Inflationary models explain the origin of density fluctuations that later led to the formation of galaxies and large-scale structures.

5.6.4 Real-world Impact:

Cosmic inflation provides a compelling framework for understanding the large-scale structure and uniformity of the universe.

5.6.5 Inflationary Models:

Slow-Roll Inflation:

In slow-roll inflation, the inflaton field rolls slowly down its potential, leading to a prolonged period of accelerated expansion.

Chaotic Inflation:

Chaotic inflation models involve a scalar field starting far from the minimum of its potential, resulting in a chaotic and rapid expansion.

5.7 Future of the Universe

Grasp the Future of the Universe with simplicity:

5.7.1 Dark Energy Dominance:

As the universe expands, dark energy becomes increasingly dominant, causing accelerated expansion.

5.7.2 Mathematical Formulas:

Hubble Parameter Evolution:

The evolution of the Hubble parameter (H) in a dark energy-dominated universe is approximated by:

$$H(a) = H_0 \sqrt{\Omega_\Lambda + \Omega_k a^{-2} + \Omega_m a^{-3}}$$

Where H_0 is the present-day Hubble constant, Ω_Λ is the dark energy density parameter, Ω_k is the curvature parameter, Ω_m is the matter density parameter, and a is the scale factor.

5.7.3   Observable Consequences:

Redshift-Distance Relationship:

The relationship between redshift (z) and distance (d) evolves, impacting our ability to observe distant galaxies:

$$v = H_0 \int_0^z \frac{dz'}{\sqrt{\Omega_\Lambda + \Omega_k(1 + z')^2 + \Omega_m(1 + z')^3}}$$

Where v is the velocity of a galaxy at redshift z.

5.7.4   Cosmic Fate Scenarios:

Open Universe:

If $\Omega_k > 0$, the universe is open, and expansion continues indefinitely.

Closed Universe:

If $\Omega_k < 0$, the universe is closed, eventually leading to a cosmic contraction.

Flat Universe:

A flat universe $(\Omega_k = 0)$ approaches a state of perpetual, but decelerating, expansion.

5.7.5   Ultimate Fate:

The ultimate fate of the universe depends on the balance between dark energy, dark matter, and other components.

5.7.6   Heat Death Scenario:

In the absence of a significant change, the universe may approach a state of heat death, with low energy and maximum entropy.

5.7.7 Real-world Impact:

Understanding the future of the universe informs our broader understanding of fundamental physics and the fate of cosmic structures.

Chapter 6

High-Energy Astrophysics

6.1 Black Holes

Grasp Black Holes with simplicity:

6.1.1 Definition:

A black hole is a region of spacetime where gravity is so strong that nothing, not even light, can escape from it.

6.1.2 Mathematical Formulas:

Schwarzschild Radius:

The Schwarzschild radius (r_s) is the critical radius defining the boundary of a non-rotating black hole:

$$r_s = \frac{2GM}{c^2}$$

Where G is the gravitational constant, M is the mass of the black hole, and c is the speed of light.

Event Horizon:

The event horizon is the boundary surrounding a black hole. For a non-rotating black hole:

$$r_{\text{horizon}} = r_s$$

Escape Velocity:

The escape velocity near the event horizon is equal to the speed of light:

$$v_{\text{esc}} = c$$

6.1.3 Observable Consequences:

Gravitational Lensing:

Black holes cause gravitational lensing, bending the path of light around them.

Accretion Disks:

Accretion disks form around black holes, emitting X-rays and other high-energy radiation.

6.1.4 Types of Black Holes:

Stellar Black Holes:

Formed from the gravitational collapse of massive stars, typically ranging from a few to tens of solar masses.

Supermassive Black Holes:

Found at the centers of galaxies, with masses ranging from hundreds of thousands to billions of solar masses.

6.1.5 Real-world Impact:

Understanding black holes is crucial for unraveling the mysteries of extreme gravitational environments in the universe.

6.1.6 Hawking Radiation:

Stephen Hawking proposed that black holes can emit radiation due to quantum effects near the event horizon, known as Hawking radiation.

6.2 Neutron Stars

Unlock the mysteries of Neutron Stars effortlessly:

6.2.1 Marvelous Neutron Star Facts:

- Neutron stars are remnants of massive stars, packing incredible density and composed mostly of neutrons.

- Resulting from supernova explosions, neutron stars exhibit fascinating properties under extreme gravity.

6.2.2 Quick Mathematical Insights:

Mass-Radius Relation (TOV Equations):

$$\frac{dP}{dr} = -\frac{G(\varepsilon + P/c^2)(m + 4\pi r^3 P/c^2)}{r(r - 2Gm/c^2)}$$

$$\frac{dm}{dr} = 4\pi r^2 \varepsilon/c^2$$

Where P is pressure, r is the radial coordinate, G is the gravitational constant, ε is energy density, m is mass, and c is the speed of light.

Keplerian Frequency:

$$\Omega_K = \left(\frac{GM}{r^3}\right)^{1/2}$$

Where G is the gravitational constant, M is the mass, and r is the distance from the center.

6.2.3 Astrophysical Impacts:

Pulsars:

Rotating neutron stars with powerful magnetic fields emit beams of radiation observed as pulsars.

X-ray Emission:

Neutron stars in binary systems can accrete matter, generating intense X-ray emissions.

6.2.4   Real-world Significance:

Neutron stars offer profound insights into extreme physics, gravity, and the nature of dense matter.

6.2.5   Beyond Equations:

Equation of State challenges scientists to comprehend the mysterious interiors of neutron stars.

6.3   Gamma-Ray Bursts

Unveil the mystery of Gamma-Ray Bursts effortlessly:

6.3.1   Quick Gamma-Ray Burst Facts:

- Gamma-Ray Bursts (GRBs) are brief and intense bursts of gamma-ray radiation, often originating from distant galaxies.

- Two types: Long-duration GRBs linked to massive star collapses, and Short-duration GRBs associated with neutron star mergers.

6.3.2   Essential Mathematical Insights:

Gamma-Ray Burst Luminosity:

$$L_\gamma = 4\pi D_L^2 S_\gamma$$

Where L_γ is the luminosity, D_L is the luminosity distance, and S_γ is the observed flux.

Isotropic Energy:

$$E_{\text{iso}} = \frac{4\pi D_L^2 S_\gamma T_{90}}{(1+z)}$$

Where E_{iso} is the isotropic energy, T_{90} is the burst duration, and z is the redshift.

6.3.3   Astrophysical Impacts:

Progenitors:

GRBs provide crucial insights into massive star collapses and neutron star mergers.

Afterglows:

The study of afterglows across different wavelengths helps understand the environment surrounding GRBs.

6.3.4   Real-world Significance:

GRBs act as cosmic beacons, offering a unique perspective into the most energetic events in the universe.

6.3.5   Beyond Equations:

Unraveling the mystery of the "GRB Central Engine" remains one of the most challenging tasks in astrophysics.

6.4   Quasars and Active Galactic Nuclei

Unravel the energy-packed world of Quasars and Active Galactic Nuclei effortlessly:

6.4.1   Quick Facts:

- Quasars are extremely luminous and energetic objects at the centers of galaxies, powered by supermassive black holes.

- Active Galactic Nuclei (AGN) encompass a range of energetic phenomena, including quasars, Seyfert galaxies, and blazars.

6.4.2   Essential Insights:

Luminosity of Quasars:

$$L_{\text{quasar}} = \epsilon \dot{M} c^2$$

Where L_{quasar} is the quasar luminosity, ϵ is the accretion efficiency, $\dot{M}$ is the mass accretion rate, and c is the speed of light.

AGN Emission Mechanism:

$$P_{\text{AGN}} = \frac{GM_{\text{bh}}\dot{M}}{2R_{\text{in}}}$$

Where P_{AGN} is the AGN power, G is the gravitational constant, M_{bh} is the black hole mass, $\dot{M}$ is the mass accretion rate, and R_{in} is the innermost stable circular orbit.

6.4.3   Astrophysical Impact:

Black Hole Influence:

Quasars and AGN offer insights into the role of supermassive black holes in galactic centers.

Emission Variability:

Observing the variability of AGN emissions helps understand the dynamics near supermassive black holes.

6.4.4   Real-world Significance:

Quasars and AGN serve as cosmic laboratories, revealing the extreme conditions around supermassive black holes.

6.4.5   Beyond Equations:

The nature of the accretion process onto supermassive black holes remains a key puzzle in astrophysics.

6.5   Cosmic Rays

Dive into the dynamic world of Cosmic Rays with ease:

6.5.1 Quick Cosmic Ray Insights:

- Cosmic rays are high-energy particles originating from various astrophysical sources, including supernovae and active galactic nuclei.

- They consist of protons, electrons, and heavier nuclei, accelerated to near-light speeds.

6.5.2 Essential Mathematical Formulas:

Power-law Spectrum:

$$I(E) = I_0 \left(\frac{E}{E_0} \right)^{-\gamma}$$

Where $I(E)$ is the cosmic ray intensity at energy E, I_0 is the normalization constant, E_0 is the reference energy, and γ is the spectral index.

Larmor Radius:

$$r_L = \frac{p}{ZeB}$$

Where r_L is the Larmor radius, p is the particle momentum, Z is the atomic number, e is the elementary charge, and B is the magnetic field strength.

6.5.3 Astrophysical Impacts:

Origin Mysteries:

Unraveling the origin of cosmic rays remains one of the long-standing mysteries in astrophysics.

Galactic and Extragalactic Cosmic Rays:

Understanding the composition and acceleration mechanisms of cosmic rays in our galaxy and beyond.

6.5.4 Real-world Significance:

Cosmic rays provide a unique window into extreme astrophysical environments and accelerate scientific discovery.

6.5.5   Beyond Equations:

The study of cosmic rays connects astrophysics, particle physics, and cosmology, contributing to a broader understanding of the universe.

6.6   Particle Acceleration

Explore Particle Acceleration effortlessly:

6.6.1   Quick Particle Acceleration Insights:

- Particle acceleration is a fundamental process in astrophysics, responsible for the creation of high-energy cosmic rays and emission of energetic radiation.

- Acceleration mechanisms include shock acceleration in supernova remnants, magnetic reconnection, and acceleration by compact objects like pulsars and black holes.

6.6.2   Essential Mathematical Formulas:

Shock Acceleration - Diffusive Shock Acceleration:

$$f(p) = f_0 \left(\frac{p}{p_0} \right)^{-s} \exp\left(-\frac{p}{p_{\max}} \right)$$

Where $f(p)$ is the particle distribution function, f_0 is the normalization constant, p_0 is the reference momentum, s is the spectral index, and $p_{\max}$ is the maximum momentum.

Magnetic Reconnection Acceleration:

$$E_{\max} = e\Phi$$

Where $E_{\max}$ is the maximum energy, e is the elementary charge, and Φ is the electric potential.

6.6.3   Astrophysical Impacts:

Galactic and Extragalactic Sources:

Understanding particle acceleration contributes to unraveling the mysteries of high-energy astrophysical phenomena.

Cosmic Ray Origins:

The study of acceleration mechanisms sheds light on the origins of cosmic rays in various astrophysical environments.

6.6.4 Real-world Significance:

Particle acceleration processes are key to unlocking the energetic secrets of the universe, impacting our understanding of cosmic phenomena.

6.6.5 Beyond Equations:

Investigating the intricate details of particle acceleration bridges astrophysics and particle physics, advancing our knowledge of the cosmos.

6.7 High-Energy Astrophysical Phenomena

Uncover the mysteries of High-Energy Astrophysical Phenomena effortlessly:

6.7.1 Quick Insights:

- High-energy astrophysical phenomena encompass a wide range of energetic events and objects, from pulsars and gamma-ray bursts to quasars and active galactic nuclei.

- These phenomena involve extreme physical conditions, such as strong gravitational fields, magnetic fields, and relativistic speeds.

6.7.2 Essential Mathematical Formulas:

Relativistic Doppler Shift:

$$\frac{\lambda_{\text{obs}}}{\lambda_{\text{em}}} = \sqrt{\frac{1 + \beta}{1 - \beta}}$$

Where λ_{obs} is the observed wavelength, λ_{em} is the emitted wavelength, and β is the velocity of the emitting object in units of the speed of light.

Synchrotron Radiation Power:

$$P_{\text{sync}} \propto B^2 \gamma^2$$

Where P_{sync} is the synchrotron radiation power, B is the magnetic field strength, and γ is the Lorentz factor.

6.7.3   Astrophysical Impacts:

Multimessenger Astronomy:

High-energy phenomena contribute to multimessenger astronomy, combining observations of different cosmic messengers, such as light, gravitational waves, and neutrinos.

Black Hole Dynamics:

Studying high-energy processes near black holes provides insights into the dynamics of spacetime.

6.7.4   Real-world Significance:

Understanding high-energy astrophysical phenomena expands our knowledge of the most extreme conditions in the universe.

6.7.5   Beyond Equations:

Ongoing research strives to unravel the connections between different high-energy phenomena, creating a comprehensive picture of the cosmos.

Chapter 7

Observational Techniques

7.1 Telescopes and Observatories

Unlock the secrets of Telescopes and Observatories effortlessly:

7.1.1 Quick Insights:

- Telescopes are vital tools for observing celestial objects, ranging from visible light to other parts of the electromagnetic spectrum.

- Observatories, equipped with various instruments, enable astronomers to capture and analyze cosmic phenomena.

7.1.2 Essential Mathematical Formulas:

Angular Resolution:

$$\theta = \frac{\lambda}{D}$$

Where θ is the angular resolution, λ is the wavelength of light, and D is the diameter of the telescope.

Light Gathering Power:

$$LGP \propto D^2$$

Where LGP is the light gathering power, and D is the diameter of the telescope.

7.1.3 Practical Considerations:

Aperture Size:

Larger telescopes with increased aperture size collect more light, enhancing their observational capabilities.

Observing Conditions:

Clear and stable atmospheric conditions significantly impact the quality of observations.

7.1.4 Real-world Significance:

Telescopes and observatories play a pivotal role in advancing our understanding of the universe, from nearby planets to distant galaxies.

7.1.5 Beyond Equations:

Technological advancements continue to enhance observational capabilities, opening new frontiers in astronomy.

7.2 Spectroscopy

Unveil the world of Spectroscopy effortlessly:

7.2.1 Quick Insights:

- Spectroscopy is a powerful technique for studying the composition, temperature, density, and motion of celestial objects.

- It involves the analysis of the interaction between light and matter, producing spectra that reveal unique signatures.

7.2.2 Essential Mathematical Formulas:

Doppler Shift for Radial Velocity:

$$\frac{\Delta\lambda}{\lambda_0} = \frac{v}{c}$$

Where $\Delta\lambda$ is the change in wavelength, λ_0 is the rest wavelength, v is the radial velocity, and c is the speed of light.

Spectral Resolution:

$$\text{Resolution} = \frac{\lambda}{\Delta\lambda}$$

Where λ is the wavelength, and $\Delta\lambda$ is the smallest detectable wavelength difference.

7.2.3 Practical Considerations:

Instrumentation:

High-resolution spectrographs and detectors enhance the accuracy of spectral measurements.

Observational Modes:

Different observational modes, such as absorption and emission spectroscopy, provide unique insights into celestial objects.

7.2.4 Real-world Significance:

Spectroscopy enables astronomers to determine the chemical composition, temperature, and other crucial properties of distant objects.

7.2.5 Beyond Equations:

Advancements in spectroscopic techniques contribute to groundbreaking discoveries, from exoplanet atmospheres to the distant universe.

7.3 Photometry

Illuminate the essence of Photometry effortlessly:

7.3.1 Quick Insights:

- Photometry is the measurement of light intensity, often used to determine the brightness of celestial objects.

- It involves the analysis of the flux or number of photons received from an astronomical source.

7.3.2   Essential Mathematical Formulas:

Flux and Luminosity:

$$F = \frac{L}{4\pi d^2}$$

Where F is the flux, L is the luminosity, and d is the distance to the object.

Apparent Magnitude:

$$m = -2.5 \log_{10}\left(\frac{F}{F_0}\right)$$

Where m is the apparent magnitude, F is the flux of the object, and F_0 is a reference flux.

7.3.3   Practical Considerations:

Filter Systems:

Photometric observations often use specific filter systems (e.g., UBVRI) to capture light in defined wavelength ranges.

Calibration Techniques:

Precise calibration of photometric instruments ensures accurate measurements and comparisons.

7.3.4   Real-world Significance:

Photometry serves as a fundamental tool for understanding the properties and variability of astronomical objects.

7.3.5   Beyond Equations:

Ongoing advancements in photometric technology contribute to the discovery and characterization of various celestial phenomena.

7.4   Radio Astronomy

Tune into the world of Radio Astronomy effortlessly:

7.4.1 Quick Insights:

- Radio Astronomy involves the study of celestial objects using radio waves, providing unique insights into the universe.

- Radio telescopes capture and analyze radio frequency emissions from various astronomical sources.

7.4.2 Essential Mathematical Formulas:

Radio Flux Density:

$$S_\nu = \frac{2kT_{\text{sys}}}{A_{\text{eff}}} \sqrt{\frac{\Delta\nu}{\tau}}$$

Where S_ν is the radio flux density, k is the Boltzmann constant, T_{sys} is the system temperature, A_{eff} is the effective area of the telescope, $\Delta\nu$ is the bandwidth, and τ is the integration time.

Radio Interferometry:

$$\text{Resolution} \propto \frac{\lambda}{B}$$

Where λ is the wavelength of observation, and B is the baseline length between radio telescopes.

7.4.3 Practical Considerations:

Radio Telescope Arrays:

Interconnected arrays of radio telescopes enhance sensitivity and resolution.

Observing Frequencies:

Different frequencies reveal unique information about celestial sources.

7.4.4 Real-world Significance:

Radio Astronomy unveils hidden aspects of the cosmos, from pulsars and quasars to the cosmic microwave background.

7.4.5   Beyond Equations:

Continual advancements in radio telescope technology expand our understanding of the universe, fostering new discoveries.

7.5   Space Observatories

Embark on the realm of Space Observatories effortlessly:

7.5.1   Quick Insights:

- Space observatories are astronomical instruments stationed in space, free from atmospheric interference, enabling unprecedented observations.

- They capture various wavelengths, from visible light to gamma rays, providing a comprehensive view of the cosmos.

7.5.2   Essential Mathematical Formulas:

Angular Resolution:

$$\theta = \frac{\lambda}{D}$$

Where θ is the angular resolution, λ is the wavelength of light, and D is the diameter of the telescope.

Spectral Resolution:

$$\text{Resolution} = \frac{\lambda}{\Delta\lambda}$$

Where λ is the wavelength, and $\Delta\lambda$ is the smallest detectable wavelength difference.

7.5.3   Practical Considerations:

Instrumentation Diversity:

Space observatories house a variety of instruments, including spectrographs, cameras, and detectors, to capture diverse astronomical data.

Orbital Characteristics:

Precise orbital parameters optimize observations, avoiding interference from Earth and other celestial bodies.

7.5.4 Real-world Significance:

Space observatories, like the Hubble Space Telescope, revolutionize our understanding of the universe, unveiling distant galaxies and cosmic phenomena.

7.5.5 Beyond Equations:

Ongoing missions and the development of new space observatories promise continuous revelations, expanding the frontiers of space exploration.

7.6 Interferometry

Merge insights effortlessly into the world of Interferometry:

7.6.1 Quick Insights:

- Interferometry combines signals from multiple telescopes to enhance resolution and extract fine details of celestial objects.

- This technique is widely used in radio, optical, and infrared astronomy to achieve high angular resolution.

7.6.2 Essential Mathematical Formulas:

Visibility Function:

$$V(u,v) = \int \int I(x,y) e^{-2\pi i(ux+vy)} \, dx \, dy$$

Where $V(u,v)$ is the visibility function, $I(x,y)$ is the brightness distribution of the source, and (u,v) are the spatial frequencies.

Interferometer Baseline:

$$B = \sqrt{u^2 + v^2}$$

Where B is the baseline length, and (u, v) are the spatial frequencies.

7.6.3 Practical Considerations:

Array Configurations:

Different configurations of telescopes in an interferometric array impact the achievable resolution and sensitivity.

Data Reduction Techniques:

Advanced algorithms process interferometric data to reconstruct images and extract scientific information.

7.6.4 Real-world Significance:

Interferometry enables astronomers to study fine details of astronomical objects, from protostellar disks to distant quasars.

7.6.5 Beyond Equations:

The expansion of interferometric arrays, like the Event Horizon Telescope, opens new possibilities for observing black holes and other astrophysical phenomena.

7.7 Data Analysis and Interpretation

Decode the essence of Data Analysis and Interpretation effortlessly:

7.7.1 Quick Insights:

- Data analysis is a critical step in turning raw observational data into meaningful scientific results.

- Interpretation involves extracting valuable information from datasets to understand the underlying astrophysical processes.

7.7.2 Essential Mathematical Formulas:

Statistical Analysis:

$$\text{Mean}(\mu) = \frac{1}{N} \sum_{i=1}^{N} x_i$$

$$\text{Standard Deviation}(\sigma) = \sqrt{\frac{1}{N} \sum_{i=1}^{N} (x_i - \mu)^2}$$

Where N is the number of data points, and x_i is each individual data point.

Curve Fitting:

Fitting observational data to mathematical models enhances our understanding of physical phenomena.

7.7.3 Practical Considerations:

Instrumental Effects:

Correcting for instrumental biases and calibration errors is crucial for accurate data analysis.

Model Selection:

Choosing appropriate astrophysical models for interpretation based on the nature of the observed phenomena.

7.7.4 Real-world Significance:

Data analysis and interpretation drive discoveries, from characterizing exoplanets to understanding the evolution of galaxies.

7.7.5 Beyond Equations:

Advancements in machine learning and artificial intelligence contribute to more sophisticated data analysis techniques in modern astrophysics.

Chapter 8

Exoplanets and Astrobiology

8.1 Methods of Detection

Uncover the secrets of Exoplanet Detection effortlessly:

8.1.1 Quick Insights:

- Detecting exoplanets is a crucial step in understanding the diversity of planetary systems beyond our solar system.

- Various methods are employed, each offering unique advantages and challenges.

8.1.2 Essential Mathematical Formulas:

Radial Velocity Method:

$$v_r = v_0 + v \sin(i) \cos(\omega + \phi)$$

Where v_r is the radial velocity, v_0 is the systemic velocity, v is the amplitude of the velocity variations, i is the orbital inclination, ω is the longitude of periastron, and ϕ is the orbital phase.

Transit Method:

$$\frac{\Delta F}{F} = \left(\frac{R_p}{R_*} \right)^2$$

Where ΔF is the change in flux, F is the stellar flux, R_p is the planet radius, and R_* is the stellar radius.

8.1.3   Practical Considerations:

Observational Cadence:

Optimizing the timing and frequency of observations enhances the probability of detection.

Follow-up Spectroscopy:

Confirming exoplanet detections often involves follow-up spectroscopic observations.

8.1.4   Real-world Significance:

Exoplanet detection has revealed a vast array of planetary systems, from hot Jupiters to Earth-like exoplanets.

8.1.5   Beyond Equations:

Cutting-edge missions like the Kepler Space Telescope and TESS (Transiting Exoplanet Survey Satellite) continue to revolutionize exoplanet detection and characterization.

8.2   Characterization of Exoplanets

Unveil the essence of Exoplanet Characterization effortlessly:

8.2.1   Quick Insights:

- Characterizing exoplanets involves determining key properties like mass, radius, and atmospheric composition.

- Different methods and instruments contribute to unraveling the mysteries of exoplanetary atmospheres.

8.2.2   Essential Mathematical Formulas:

Mass-Radius Relationship:

$$\frac{M_p}{M_J} = \left(\frac{R_p}{R_J}\right)^{\alpha}$$

Where M_p is the exoplanet mass, M_J is Jupiter's mass, R_p is the exoplanet radius, R_J is Jupiter's radius, and α is a scaling exponent.

Transit Spectroscopy:

$$\frac{\Delta R_p}{R_p} \propto \frac{\Delta F}{F}$$

Where ΔR_p is the change in radius, R_p is the exoplanet radius, ΔF is the change in flux, and F is the stellar flux.

8.2.3 Practical Considerations:

Spectral Analysis:

Extracting atmospheric composition through spectroscopic analysis of transiting exoplanets.

Phase Curves:

Studying the variations in an exoplanet's brightness over its orbit provides insights into its atmosphere.

8.2.4 Real-world Significance:

Characterizing exoplanets unveils a diverse range of atmospheres, from hot and gaseous to potentially habitable.

8.2.5 Beyond Equations:

Future space missions, like the James Webb Space Telescope, promise unprecedented capabilities for exoplanet characterization.

8.3 Habitability

Explore the realm of Habitability effortlessly:

8.3.1 Quick Insights:

- Habitability assesses the conditions under which a planet could support life as we know it.

- Key factors include the planet's distance from its star, atmospheric composition, and surface conditions.

8.3.2   Essential Mathematical Formulas:

Habitable Zone:

$$S = S_0 \sqrt{\frac{L}{L_0}}$$

Where S is the flux received by the planet, S_0 is the solar constant, L is the star's luminosity, and L_0 is the solar luminosity.

Surface Temperature:

$$T_s = T_{\text{eff}} \sqrt{\frac{R_*}{2D}}$$

Where T_s is the equilibrium temperature, T_{eff} is the effective temperature of the star, R_* is the star's radius, and D is the distance from the star.

8.3.3   Practical Considerations:

Atmospheric Conditions:

Assessing the composition and stability of an exoplanet's atmosphere is crucial for habitability.

Liquid Water:

Considering the presence of liquid water, a key ingredient for life as we know it.

8.3.4   Real-world Significance:

Understanding habitability guides the search for potentially life-supporting exoplanets within and beyond our galaxy.

8.3.5   Beyond Equations:

Ongoing and future missions, like the LUVOIR (Large UV/Optical/IR Surveyor) telescope, aim to study exoplanet atmospheres and assess their habitability.

8.4   Search for Extraterrestrial Life

Embark on the quest for Extraterrestrial Life effortlessly:

8.4.1 Quick Insights:

- The search for extraterrestrial life involves scanning exoplanets for biosignatures and intelligent signals.

- Diverse methods and technologies contribute to this captivating endeavor.

8.4.2 Essential Mathematical Formulas:

Habitable Zone Estimate:

$$D_{\mathrm{HZ}} = \sqrt{\frac{L}{L_0}}$$

Where D_{HZ} is the estimated Habitable Zone distance, L is the star's luminosity, and L_0 is the solar luminosity.

Drake Equation:

$$N = R^* \cdot f_p \cdot n_e \cdot f_\ell \cdot f_i \cdot f_c \cdot L$$

Where N is the estimated number of civilizations, and the factors represent various parameters influencing the probability of extraterrestrial life.

8.4.3 Practical Considerations:

Technological Signatures:

Searching for artificial signals, such as radio waves, as potential indicators of advanced civilizations.

Biosignature Detection:

Identifying chemical signatures in exoplanet atmospheres that could indicate the presence of life.

8.4.4 Real-world Significance:

The search for extraterrestrial life captivates scientists and the public alike, influencing our understanding of our place in the universe.

8.4.5 Beyond Equations:

Advancements in telescope technology, like the James Webb Space Telescope, expand our capabilities to study potentially habitable exoplanets.

8.5 Drake Equation

Uncover the mysteries of the Drake Equation effortlessly:

8.5.1 Quick Insights:

- The Drake Equation estimates the potential number of extraterrestrial civilizations in our galaxy.

- It considers various factors that influence the probability of advanced civilizations.

8.5.2 Essential Mathematical Formulas:

$$N = R^* \cdot f_p \cdot n_e \cdot f_\ell \cdot f_i \cdot f_c \cdot L$$

Where:

N is the estimated number of civilizations,

R^* is the rate of star formation in our galaxy,

f_p is the fraction of stars with planetary systems,

n_e is the average number of planets that could support life per star with planets,

f_ℓ is the fraction of those planets where life actually develops,

f_i is the fraction of planets with intelligent life,

f_c is the fraction of planets with technological civilizations,

L is the average lifespan of such civilizations.

8.5.3 Practical Considerations:

- The Drake Equation sparks discussions about the factors influencing the existence of extraterrestrial intelligence.

- Each factor involves scientific and philosophical considerations, making it a tool for contemplation.

8.5.4 Real-world Significance:

- The Drake Equation encourages interdisciplinary thinking and serves as a framework for discussions about the search for extraterrestrial life.

8.5.5 Beyond Equations:

- The ongoing quest to refine the parameters of the Drake Equation reflects our evolving understanding of the cosmos.

8.6 Astrobiology in the Solar System

Unveil the wonders of Astrobiology in our Solar System effortlessly:

8.6.1 Quick Insights:

- Astrobiology explores the potential for life beyond Earth, including our Solar System's celestial bodies.

- Key targets for astrobiological studies include Mars, Europa, Enceladus, and Titan.

8.6.2 Essential Mathematical Formulas:

Mars Exploration:

$$\Delta v = v_{\text{escape}} \sqrt{\frac{2r}{r + h}}$$

Where:

Δv is the required velocity change for a spacecraft,

v_{escape} is the escape velocity from Mars,

r is the initial orbital radius,

h is the altitude of the target orbit.

Europa Exploration:

$$P = P_0 e^{-\frac{h}{H}}$$

Where:

P is the atmospheric pressure,

P_0 is the atmospheric pressure at the surface,

h is the altitude,

H is the scale height of the atmosphere.

8.6.3 Practical Considerations:

Mars Surface Conditions:

Understanding the challenges and opportunities for potential life on Mars based on atmospheric and surface conditions.

Ocean Worlds Exploration:

Investigating subsurface oceans on moons like Europa and Enceladus for signs of life.

8.6.4 Real-world Significance:

- Astrobiology in our Solar System informs future exploration missions and shapes our understanding of the potential for life beyond Earth.

8.6.5 Beyond Equations:

- Ongoing and planned missions, such as the Europa Clipper, aim to explore astrobiologically intriguing environments.

8.7 Exoplanetary Atmospheres

Explore the mysteries of Exoplanetary Atmospheres effortlessly:

8.7.1 Quick Insights:

- Exoplanetary atmospheres play a crucial role in determining a planet's habitability and potential for life.

- Key components include composition, temperature, and pressure.

8.7.2 Essential Mathematical Formulas:

Ideal Gas Law:

$$PV = nRT$$

Where:

P is the pressure of the gas,

V is the volume,

n is the number of moles of gas,

R is the ideal gas constant,

T is the temperature.

Escape Velocity:

$$v_{\text{escape}} = \sqrt{\frac{2GM}{R}}$$

Where:

v_{escape} is the escape velocity,

G is the gravitational constant,

M is the planet's mass,

R is the planet's radius.

8.7.3 Practical Considerations:

Composition Analysis:

Studying the spectroscopic signatures of exoplanetary atmospheres to determine their chemical composition.

Habitability Assessment:

Assessing the potential habitability of exoplanets based on atmospheric conditions.

8.7.4   Real-world Significance:

- Understanding exoplanetary atmospheres informs our search for habitable exoplanets and extraterrestrial life.

8.7.5   Beyond Equations:

- Advances in observational techniques, such as transmission spectroscopy, enhance our ability to study exoplanetary atmospheres.

Chapter 9

High-Performance Computing in Astrophysics

9.1 Simulation Techniques

Dive into the world of Simulation Techniques with ease:

9.1.1 Quick Insights:

- Simulation techniques in astrophysics enable the modeling of complex phenomena, from stellar evolution to galaxy formation.

- Key components include numerical methods, algorithms, and high-performance computing (HPC) resources.

9.1.2 Essential Mathematical Formulas:

N-Body Simulation:

The gravitational force between two bodies:

$$F = \frac{Gm_1m_2}{r^2}$$

97

Where:

F is the gravitational force,

G is the gravitational constant,

m_1, m_2 are the masses of the two bodies,

r is the separation between the two bodies.

Fluid Dynamics Simulation:

Navier-Stokes equations for fluid flow:

$$\frac{\partial \rho}{\partial t} + \nabla \cdot (\rho \mathbf{v}) = 0$$

$$\rho \left(\frac{\partial \mathbf{v}}{\partial t} + (\mathbf{v} \cdot \nabla)\mathbf{v} \right) = -\nabla P + \nabla \cdot \o + \rho \mathbf{g}$$

Where:

ρ is the density of the fluid,

$\mathbf{v}$ is the velocity vector field,

P is the pressure,

$\o$ is the stress tensor,

$\mathbf{g}$ is the gravitational acceleration vector.

9.1.3   Practical Considerations:

Parallel Computing:

Optimizing simulations for parallel processing on HPC architectures.

Adaptive Mesh Refinement:

Refining computational grids dynamically to focus resources on regions of interest.

9.1.4   Real-world Significance:

- Simulation techniques drive advancements in understanding astrophysical phenomena, guiding observational efforts and theoretical developments.

9.1.5 Beyond Equations:

- High-performance computing facilities, such as supercomputers, empower simulations that were once considered computationally infeasible.

9.2 Numerical Methods

Explore the realm of Numerical Methods effortlessly:

9.2.1 Quick Insights:

- Numerical methods are essential for solving complex astrophysical equations that lack analytical solutions.

- Key techniques include finite difference, finite element, and spectral methods.

9.2.2 Essential Mathematical Formulas:

Finite Difference Method:

Approximating derivatives using finite differences:

$$f'(x) \approx \frac{f(x+h) - f(x)}{h}$$

Where:

$f'(x)$ is the derivative of $f(x)$,

h is the step size.

Finite Element Method:

Discretizing differential equations over finite elements:

$$\mathbf{Ku} = \mathbf{f}$$

Where:

$$\mathbf{K} \text{ is the stiffness matrix,}$$

$$\mathbf{u} \text{ is the solution vector,}$$

$$\mathbf{f} \text{ is the forcing vector.}$$

9.2.3 Practical Considerations:

Accuracy vs. Computational Cost:

Balancing the trade-off between accuracy and computational efficiency in numerical simulations.

Convergence Analysis:

Assessing the convergence of numerical solutions to ensure accuracy and reliability.

9.2.4 Real-world Significance:

- Numerical methods enable the modeling of diverse astrophysical phenomena, from fluid dynamics in stars to gravitational interactions between galaxies.

9.2.5 Beyond Equations:

- High-performance computing platforms empower the implementation of numerical methods for large-scale simulations, pushing the boundaries of astrophysical research.

9.3 Parallel Computing

Embark on the world of Parallel Computing in Astrophysics effortlessly:

9.3.1 Quick Insights:

- Parallel computing is the key to harnessing the full potential of high-performance computing (HPC) resources in astrophysical simulations.

- It involves breaking down complex problems into smaller tasks that can be solved simultaneously.

9.3.2 Essential Mathematical Formulas:

Speedup in Parallel Computing:

The speedup (S) achieved using P processors:

$$S = \frac{T_1}{T_P}$$

Where:

T_1 is the execution time with a single processor,

T_P is the execution time with P processors.

Efficiency of Parallel Computing:

Efficiency (E) of using P processors:

$$E = \frac{S}{P}$$

Where:

S is the speedup,

P is the number of processors.

9.3.3 Practical Considerations:

Load Balancing:

Distributing the computational workload evenly among processors for optimal performance.

Communication Overhead:

Minimizing the time spent on inter-processor communication.

9.3.4 Real-world Significance:

- Parallel computing allows astrophysicists to tackle more extensive and complex simulations, advancing our understanding of the cosmos.

9.3.5   Beyond Equations:

- Implementation of parallel algorithms on HPC architectures accelerates the pace of astrophysical discoveries.

9.4   Data Visualization

Immerse yourself in the realm of Data Visualization in Astrophysics effortlessly:

9.4.1   Quick Insights:

- Data visualization is the art of representing complex astrophysical datasets in a visually intuitive manner.

- It aids in extracting meaningful insights from large-scale simulations and observations.

9.4.2   Practical Techniques:

Color Mapping:

Assigning colors to represent different parameters enhances the understanding of multidimensional datasets.

3D Rendering:

Transforming numerical data into three-dimensional visualizations for a more immersive experience.

9.4.3   Essential Mathematical Formulas:

Histograms for Data Distribution:

Dividing data into bins and visualizing the frequency distribution:

$$P(x) = \frac{\text{Number of data points in bin } x}{\text{Total number of data points}}$$

Heatmaps for Correlation:

Visualizing the correlation between two variables:

$$\text{Correlation} = \frac{\text{Covariance}(X, Y)}{\text{Standard Deviation}(X) \times \text{Standard Deviation}(Y)}$$

9.4.4 Practical Considerations:

Interactive Visualization:

Allowing scientists to interact with and explore the data dynamically.

Virtual Reality (VR) Visualization:

Taking data visualization to the next level with immersive VR experiences.

9.4.5 Real-world Significance:

- Data visualization facilitates the communication of astrophysical findings to both experts and the general public.

9.4.6 Beyond Equations:

- Advanced visualization tools contribute to a deeper understanding of celestial phenomena, from galaxies to cosmological structures.

9.5 Computational Challenges

Embark on the exploration of Computational Challenges in Astrophysics effortlessly:

9.5.1 Quick Insights:

- Astrophysical simulations face unique computational challenges due to the vast scale and complexity of celestial phenomena.

- Overcoming these challenges is essential for advancing our understanding of the universe.

9.5.2   Key Challenges:

Numerical Instabilities:

Addressing issues arising from the discretization of continuous equations:

$$\text{Stability} \propto \frac{\Delta t}{\Delta x^2}$$

Parallelization Overhead:

Balancing load distribution among processors for optimal parallel performance:

$$\text{Parallel Efficiency} = \frac{\text{Speedup}}{\text{Number of Processors}}$$

9.5.3   Practical Considerations:

Adaptive Mesh Refinement:

Adjusting the computational grid dynamically to focus resources on regions of interest.

Algorithmic Optimization:

Fine-tuning numerical methods for efficiency in specific astrophysical scenarios.

9.5.4   Real-world Significance:

- Overcoming computational challenges enables accurate simulations of cosmic events, from stellar explosions to galaxy formations.

9.5.5   Beyond Equations:

- Collaborative efforts in algorithm development and hardware advancements play a pivotal role in conquering computational challenges in astrophysical simulations.

9.6   Code Optimization

Embark on the journey of Code Optimization in Astrophysics effortlessly:

9.6.1 Quick Insights:

- Code optimization is crucial for maximizing the efficiency of astrophysical simulations and computations.

- Well-optimized code leads to faster results, enabling more extensive and detailed simulations.

9.6.2 Key Optimization Techniques:

Loop Unrolling:

Expanding loops to reduce overhead and improve parallelization:

$$\text{Optimized Loop}: \quad \text{for}(i = 0; i < N; i+ = 2)\{\ldots\}$$

Vectorization:

Utilizing SIMD (Single Instruction, Multiple Data) instructions for parallel processing:

$$\text{Vectorized Operation}: \quad \text{result} = \text{vector\_A} \times \text{vector\_B}$$

9.6.3 Practical Considerations:

Memory Access Patterns:

Optimizing data structures and access patterns to minimize cache misses.

Algorithmic Complexity:

Choosing algorithms with lower time complexity for specific astrophysical computations.

9.6.4 Real-world Significance:

- Well-optimized code allows researchers to simulate complex astrophysical phenomena more efficiently, from galaxy mergers to gravitational wave events.

9.6.5 Beyond Equations:

- Collaborative efforts in code optimization contribute to the development of high-performance astrophysical simulations, pushing the boundaries of computational capabilities.

9.7 Future Trends

Explore the Future Trends in High-Performance Computing in Astrophysics effortlessly:

9.7.1 Quick Insights:

- Anticipating the future of high-performance computing (HPC) in astrophysics is essential for pushing the boundaries of scientific exploration.

- Innovations in hardware and algorithms are key drivers shaping the future landscape of astrophysical simulations.

9.7.2 Trend 1: Quantum Computing Impact

Quantum Advantage:

Harnessing quantum computing for solving complex astrophysical problems:

$$\text{Quantum Speedup} \propto 2^n \quad \text{(for certain problems)}$$

Quantum Algorithms:

Developing quantum algorithms for simulating quantum systems, such as many-body celestial interactions.

9.7.3 Trend 2: Exascale Computing

Exascale Systems:

Exploiting the power of exascale computing for unprecedented simulation capabilities:

$$\text{Exaflop} = 10^{18} \text{ floating-point operations per second}$$

Parallelization Strategies:

Adopting advanced parallelization techniques to fully leverage exascale architectures.

9.7.4 Trend 3: Machine Learning Integration

Deep Learning in Astrophysics:

Integrating deep learning for data analysis, pattern recognition, and optimization:

$$\text{Neural Network Output} = \sigma(\text{Weighted Sum of Inputs})$$

Automated Parameter Tuning:

Utilizing machine learning to optimize simulation parameters for efficiency.

9.7.5 Real-world Significance:

- Embracing these future trends promises groundbreaking advancements, enabling more accurate and detailed simulations of cosmic phenomena.

9.7.6 Beyond Equations:

- Collaborations between astrophysicists, computer scientists, and hardware engineers drive the implementation of these trends, shaping the future of computational astrophysics.

Chapter 10

Astrophysical Instrumentation

10.1 Detectors

Unveil the world of Detectors in Astrophysical Instrumentation effortlessly:

10.1.1 Quick Insights:

- Detectors play a pivotal role in capturing and analyzing celestial signals across the electromagnetic spectrum.

- Understanding the principles behind detectors is essential for interpreting astronomical observations accurately.

10.1.2 Detector Types:

Photodetectors:

Efficiently convert photons into electrical signals:

$$\text{Quantum Efficiency} = \frac{\text{Number of Photons Detected}}{\text{Number of Photons Incident}}$$

Charge-Coupled Devices (CCDs):

Utilized for precise imaging and spectroscopy:

$$\text{Signal-to-Noise Ratio (SNR)} = \frac{\text{Signal Strength}}{\text{Noise Level}}$$

10.1.3 Practical Considerations:

Dark Current:

Minimizing thermal noise in detectors:

$$\text{Dark Current} \propto \exp\left(\frac{-E_{\text{activation}}}{kT}\right)$$

Readout Noise:

Optimizing readout processes to reduce electronic noise.

10.1.4 Detector Calibration:

Flat Fielding:

Compensating for pixel-to-pixel variations:

$$\text{Calibrated Intensity} = \frac{\text{Raw Intensity}}{\text{Flat Field}}$$

Bias Subtraction:

Removing electronic offset for accurate data interpretation.

10.1.5 Real-world Significance:

- Detector advancements enhance the precision and reliability of astronomical observations, from imaging distant galaxies to studying exoplanet atmospheres.

10.1.6 Beyond Equations:

- Collaborations between astronomers and engineers drive innovation in detector technologies, opening new frontiers in observational astronomy.

10.2 Spectrometers

Embark on a quick journey into Spectrometers in Astrophysical Instrumentation:

10.2.1 Snapshot of Spectrometers:

- Spectrometers are vital tools for dissecting the light from celestial objects, unraveling their composition and physical properties.

- Understanding the basic principles of spectrometry enhances our ability to decode the secrets of the universe.

10.2.2 Key Components:

Grating or Prism:

$$\text{Dispersion} = \frac{\text{Change in Angle}}{\text{Change in Wavelength}}$$

Detector Array:

$$\text{Spectral Resolution} = \frac{\lambda}{\Delta\lambda}$$

10.2.3 Modes of Operation:

Imaging Spectroscopy:

Captures a spectrum at each spatial position:

$$\text{2D Spectrum} = f(\text{Spatial Position}, \lambda)$$

Integral Field Spectroscopy:

Simultaneously obtains spectra across an extended field.

10.2.4 Calibration Techniques:

Wavelength Calibration:

Aligning spectral features with known wavelengths:

$$\text{Accuracy} = \frac{\text{Observed Wavelength}}{\text{True Wavelength}}$$

Flux Calibration:

Determining the energy output at each wavelength.

10.2.5   Practical Significance:

- Spectrometers enable astronomers to unravel the chemical composition, temperature, and velocity of celestial objects.

10.2.6   Real-world Application:

- From characterizing exoplanet atmospheres to studying distant galaxies, spectrometers play a pivotal role in expanding our cosmic understanding.

10.3   Cameras

Embark on a quick exploration of Cameras in Astrophysical Instrumentation:

10.3.1   Insights into Cameras:

- Cameras serve as the eyes of astronomers, capturing the beauty and mysteries of the cosmos.

- Understanding the basics of astronomical cameras is crucial for interpreting celestial images.

10.3.2   Key Components:

Charge-Coupled Device (CCD):

$$\text{Signal-to-Noise Ratio (SNR)} = \frac{\text{Signal}}{\text{Noise}}$$

Quantum Efficiency:

$$\text{Quantum Efficiency} = \frac{\text{Photons Converted to Electrons}}{\text{Incident Photons}}$$

10.3.3 Resolution Considerations:

Spatial Resolution:

$$\text{Angular Resolution} = \frac{\lambda}{D}$$

Temporal Resolution:

$$\text{Frame Rate} = \frac{1}{\text{Exposure Time}}$$

10.3.4 Noise Reduction Techniques:

Dark Current Subtraction:

$$\text{Reduced Signal} = \text{Observed Signal} - \text{Dark Current}$$

Flat Fielding:

$$\text{Corrected Image} = \frac{\text{Original Image}}{\text{Flat Field}}$$

10.3.5 Practical Significance:

- Cameras empower astronomers to capture and analyze the light from celestial objects, revealing details otherwise invisible to the human eye.

10.3.6 Real-world Application:

- From planetary exploration to deep-sky imaging, cameras contribute significantly to our understanding of the universe.

10.4 Radio Telescopes

Embark on a quick exploration of Radio Telescopes in Astrophysical Instrumentation:

10.4.1 Insights into Radio Telescopes:

- Radio Telescopes detect and analyze radio waves emitted by celestial objects, offering a unique perspective on the universe.

- Understanding the fundamental principles of radio astronomy enhances our ability to unveil cosmic mysteries.

10.4.2   Key Components:

Antenna Gain:

$$\text{Power Gain (G)} = \frac{4\pi A}{\lambda^2}$$

Receiver Sensitivity:

$$\text{Receiver Temperature (T)} = \frac{\text{Signal Power}}{\text{Noise Power}}$$

10.4.3   Resolution Considerations:

Angular Resolution:

$$\text{Angular Resolution} \approx \frac{\lambda}{D}$$

Synthesis Imaging:

$$\text{Synthesized Beam Width} \approx \frac{\lambda}{B}$$

10.4.4   Noise Reduction Techniques:

Interference Mitigation:

$$\text{Interference Reduction} = \text{Observed Signal} - \text{Interference Signal}$$

Time Averaging:

$$\text{Time-Averaged Signal} = \frac{1}{\text{Integration Time}} \int_0^{\text{Integration Time}} \text{Instantaneous Signal} \, dt$$

10.4.5   Practical Significance:

- Radio Telescopes unveil the secrets of distant galaxies, pulsars, and cosmic phenomena by capturing radio signals from space.

10.4.6 Real-world Application:

- From mapping the Milky Way's spiral structure to studying radio-emitting objects, radio telescopes play a crucial role in modern astronomy.

10.5 Space Probes and Satellites

Embark on a swift exploration of Space Probes and Satellites in Astrophysical Instrumentation:

10.5.1 Insights into Space Probes and Satellites:

- Space probes and satellites revolutionize astronomical observations, providing a unique perspective beyond Earth's atmosphere.

- Uncover the mysteries of our solar system and beyond through data collected by these advanced instruments.

10.5.2 Key Components:

Telemetry and Communication:

$$\text{Data Rate} = \frac{\text{Total Data Transmitted}}{\text{Transmission Time}}$$

Power Systems:

$$\text{Solar Panel Efficiency} = \frac{\text{Power Output}}{\text{Solar Irradiance} \times \text{Panel Area}}$$

10.5.3 Orbital Mechanics:

Orbital Velocity:

$$\text{Orbital Velocity} = \sqrt{\frac{GM}{R}}$$

Escape Velocity:

$$\text{Escape Velocity} = \sqrt{\frac{2GM}{R}}$$

10.5.4   Data Transmission:

Signal-to-Noise Ratio (SNR):

$$\text{SNR} = \frac{\text{Signal Power}}{\text{Noise Power}}$$

Link Budget:

$$\text{Received Power} = \text{Transmitted Power} + \text{Antenna Gain} - \text{Free Space Path Loss}$$

10.5.5   Practical Significance:

- Space probes and satellites enable us to explore distant planets, moons, and celestial bodies, expanding our understanding of the cosmos.

10.5.6   Real-world Application:

- From mapping exoplanets to studying the composition of asteroids, space probes and satellites play a vital role in space exploration.

10.6   Adaptive Optics

Dive into the world of Adaptive Optics with a quick and accessible overview:

10.6.1   Understanding Adaptive Optics:

- Adaptive Optics (AO) revolutionizes astronomical observations by mitigating the effects of atmospheric turbulence.

- Enhance the clarity and resolution of images captured by telescopes using adaptive mirror adjustments.

10.6.2   Key Components:

Deformable Mirrors:

$$\text{Actuator Influence Function} = \frac{\text{Change in Surface Height}}{\text{Actuator Force}}$$

Wavefront Sensor:

$$\text{Wavefront Error} = \text{Reference Wavefront} - \text{Measured Wavefront}$$

10.6.3 Correction Algorithms:

Modal Correction:

$$\text{Corrected Wavefront} = \text{Original Wavefront} - \text{Modal Coefficients}$$

Zernike Polynomials:

$$Z_n^m(\rho, \theta) = R_n^m(\rho)\cos(m\theta) \quad \text{(for even } m\text{)}$$

10.6.4 Applications in Astronomy:

- AO improves image quality, enabling detailed observations of distant galaxies, nebulae, and other celestial objects.

10.6.5 Real-world Impact:

- From planetary exploration to capturing high-resolution images of distant stars, Adaptive Optics plays a pivotal role in modern astronomy.

10.7 Instrument Calibration

Embark on the realm of Instrument Calibration with a quick and accessible overview:

10.7.1 Understanding Instrument Calibration:

- Instrument Calibration ensures the accuracy and reliability of astronomical observations.

- Precise calibration is crucial for interpreting raw data and extracting meaningful scientific information.

10.7.2   Key Components:

Calibration Models:

$$\text{Calibrated Data} = \text{Raw Data} \times \text{Calibration Factor}$$

Error Analysis:

$$\text{Total Error} = \text{Systematic Error} + \text{Random Error}$$

10.7.3   Calibration Techniques:

Flat Fielding:

$$\text{Calibrated Image} = \frac{\text{Raw Image}}{\text{Flat Field Image}}$$

Wavelength Calibration:

$$\text{Wavelength} = \text{Dispersion Relation} \times \text{Pixel Position}$$

10.7.4   Applications in Astronomy:

- Accurate instrument calibration enables astronomers to study the spectral characteristics of celestial objects with precision.

10.7.5   Real-world Impact:

- From ground-based telescopes to space observatories, instrument calibration is the cornerstone of reliable astronomical research.

Chapter 11

Astrophysical Constants and Units

11.1 Physical Constants

Dive into the world of Physical Constants with a quick and accessible overview:

11.1.1 Understanding Constants:

- Physical constants are unchanging values representing fundamental aspects of the universe.

- These constants provide a foundation for scientific calculations and theories.

11.1.2 Speed of Light:

The speed of light, denoted as c, is a fundamental constant with a value of approximately 3.00×10^8 m/s.

11.1.3 Planck's Constant:

Planck's constant, denoted as h, is a key value in quantum mechanics with a value of approximately 6.63×10^{-34} J$\cdot$s.

11.1.4 Gravitational Constant:

The gravitational constant, denoted as G, is crucial in gravitational physics with a value of approximately 6.67×10^{-11} m^3/kg$\cdot$s^2.

11.1.5   Boltzmann's Constant:

Boltzmann's constant, denoted as k, relates energy and temperature in statistical mechanics with a value of approximately 1.38×10^{-23} J/K.

11.1.6   Avogadro's Number:

Avogadro's number, denoted as N_A, represents the number of atoms or molecules in one mole and is approximately 6.02×10^{23}.

11.1.7   Unified Atomic Mass Unit:

The unified atomic mass unit, denoted as u, is used for expressing atomic and molecular masses and is approximately 1.66×10^{-27} kg.

11.1.8   Real-world Application:

- Utilized in various scientific disciplines, these constants form the backbone of astronomical calculations and experiments.

11.2   Astronomical Constants

Embark on a quick journey through essential Astronomical Constants:

11.2.1   Light-Year:

- A light-year, denoted as ly, represents the distance light travels in one year.

- It is approximately 9.46×10^{15} meters.

11.2.2   Parsec:

- A parsec, denoted as pc, is a unit of length used in astronomy.

- It is equivalent to about 3.09×10^{16} meters.

11.2.3 Solar Mass:

- The solar mass, denoted as $M_\odot$, is the mass of our Sun.

- It is approximately 1.99×10^{30} kg.

11.2.4 Astronomical Unit:

- An astronomical unit, denoted as AU, is the average distance between the Earth and the Sun.

- It is approximately 1.496×10^{11} meters.

11.2.5 Real-world Application:

- These constants provide a foundation for expressing astronomical distances, masses, and sizes in a comprehensible manner.

11.3 Unit Systems

Dive into the simplicity of Astrophysical Unit Systems:

11.3.1 SI Units:

- The International System of Units (SI) is widely used in astrophysics.

- Length is measured in meters (m), mass in kilograms (kg), and time in seconds (s).

- $\text{Energy}(J) = \text{Mass}(kg) \times (\text{Distance}(m))^2/(\text{Time}(s))^2$

11.3.2 Solar Units:

- Solar units are often employed for celestial bodies.

- Solar mass ($M_\odot$) and solar luminosity ($L_\odot$) are central:

- $L_\odot = 3.828 \times 10^{26}$ Watts

11.3.3 Parsec Units:

- Astronomers favor the parsec-based system for cosmic distances.

- Distances expressed in parsecs (pc).

11.3.4 Electron Volt:

- In particle astrophysics, energy is measured in electron volts (eV).

- $1\,\text{eV} = 1.602 \times 10^{-19}\,\text{J}$

11.3.5 Real-world Application:

- Astrophysical units simplify complex calculations, making them more accessible for researchers and students.

11.4   Conversion Factors

Embark on the simplicity of Astrophysical Conversion Factors:

11.4.1 Light-Year to Parsec:

$$1\,\text{light-year} = 0.306601\,\text{parsecs}$$

11.4.2 Parsec to Kilometers:

$$1\,\text{parsec} = 3.086 \times 10^{13}\,\text{km}$$

11.4.3 Solar Mass to Kilograms:

$$1\,M_\odot = 1.989 \times 10^{30}\,\text{kg}$$

11.4.4 Electron Volt to Joules:

$$1\,\text{eV} = 1.602 \times 10^{-19}\,\text{J}$$

11.4.5 Conversion in Celestial Mechanics:

$$F = \frac{G \cdot m_1 \cdot m_2}{r^2}$$

Where:

F is the gravitational force,

G is the gravitational constant ($6.674 \times 10^{-11}\,\mathrm{Nm^2/kg^2}$),

m_1 and m_2 are the masses,

r is the separation distance.

11.4.6 Real-world Application:

Astrophysical conversion factors facilitate seamless transitions between units, simplifying complex calculations in astronomy and astrophysics.

11.5 Dimensionless Numbers

Dive into the simplicity of Astrophysical Dimensionless Numbers:

11.5.1 Reynolds Number (Re):

$$Re = \frac{\rho \cdot v \cdot L}{\eta}$$

Where:

Re is the Reynolds number,

ρ is the density of the fluid,

v is the velocity of the fluid,

L is a characteristic length,

η is the dynamic viscosity of the fluid.

11.5.2 Mach Number (Ma):

$$Ma = \frac{v}{c}$$

Where:

Ma is the Mach number,

v is the speed of the object through the fluid,

c is the speed of sound in the fluid.

11.5.3   Prandtl Number (Pr):

$$Pr = \frac{\nu}{\alpha}$$

Where:

Pr is the Prandtl number,

ν is the kinematic viscosity,

α is the thermal diffusivity.

11.5.4   Real-world Application:

Dimensionless numbers simplify the characterization of fluid flow, heat transfer, and other physical phenomena, making them essential tools in astrophysical research.

11.6   Common Symbols

Unveil the symbols that guide the cosmos:

11.6.1   Speed of Light (c):

$$c = 3 \times 10^8 \, \text{m/s}$$

The cosmic speed limit, representing the velocity of light in a vacuum.

11.6.2   Planck's Constant (h):

$$h \approx 6.626 \times 10^{-34} \, \text{J} \cdot \text{s}$$

Quantum leap! A fundamental constant governing the energy of particles.

11.6.3 Gravitational Constant (G):

$$G \approx 6.674 \times 10^{-11}\,\mathrm{m^3 \cdot kg^{-1} \cdot s^{-2}}$$

The force behind gravity, determining the attraction between masses.

11.6.4 Boltzmann's Constant (k):

$$k \approx 1.381 \times 10^{-23}\,\mathrm{J/K}$$

Heat wisdom! Relates temperature to energy in the microscopic world.

11.6.5 Avogadro's Number (N_A):

$$N_A \approx 6.022 \times 10^{23}\,\mathrm{mol^{-1}}$$

Mole magic! The number of atoms or molecules in one mole.

11.6.6 Real-world Application:

These symbols are the cosmic alphabet, unlocking the secrets of the universe. From the speed of light to the quantum realm, they guide our understanding of astrophysical phenomena.

11.7 Useful Formulas

Unlock the universe with these astrophysical gems:

11.7.1 Energy-Mass Equivalence:

$$E = mc^2$$

Einstein's masterpiece, showcasing the interplay of energy (E) and mass (m).

11.7.2 Blackbody Radiation:

$$B(\lambda, T) = \frac{2hc^2}{\lambda^5} \cdot \frac{1}{e^{\frac{hc}{\lambda k T}} - 1}$$

Planck's law, unraveling the radiation spectrum from hot bodies.

11.7.3   Doppler Shift:

$$f' = f \left(1 + \frac{v}{c}\right)$$

The shift in frequency (f') due to the velocity (v) of a source relative to an observer.

11.7.4   Escape Velocity:

$$v_e = \sqrt{\frac{2GM}{R}}$$

The speed required to break free from a celestial body's gravitational pull.

11.7.5   Hubble's Law:

$$v = H_0 D$$

The recession velocity (v) of a galaxy is proportional to its distance (D) from us.

11.7.6   Real-world Application:

These formulas are the cosmic tools, enabling us to decipher the mysteries of the universe. From understanding radiation to predicting escape velocities, they guide us through the vastness of astrophysics.

Chapter 12

Astrophysical Data and Databases

12.1 Astronomical Catalogs

Navigate the celestial sea with these quick insights:

12.1.1 Messier Catalog:

A treasure map of deep-sky wonders, aiding astronomers in identifying and observing galaxies, nebulae, and star clusters.

12.1.2 NGC Catalog:

New General Catalog, a comprehensive compilation of celestial objects, serving as a go-to reference for astronomers.

12.1.3 HIPPARCOS Catalog:

Precision positioning of stars, offering accurate parallax measurements for a stellar cartography adventure.

12.1.4 SDSS:

Sloan Digital Sky Survey, a vast cosmic snapshot providing data on millions of galaxies, quasars, and stars.

12.1.5 Real-world Application:

Astronomical catalogs act as cosmic dictionaries, allowing astronomers to name, locate, and study celestial inhabitants efficiently.

12.2 Data Repositories

Embark on a cosmic journey with these bite-sized insights:

12.2.1 NASA Astrophysics Data System (ADS):

A treasure trove of scholarly articles, seamlessly linking researchers to a galaxy of astrophysical knowledge.

12.2.2 European Space Agency (ESA) Archives:

Home to mission data and cosmic imagery, serving astronomers and space enthusiasts with a visual feast.

12.2.3 VizieR:

Your cosmic toolbox, providing access to a wealth of astronomical catalogs and data tables for quick research.

12.2.4 Virtual Observatory (VO):

A collaborative hub where astronomers access and analyze data across various telescopes and missions.

12.2.5 Real-world Application:

Data repositories are the interstellar libraries, enabling astronomers to access, share, and unravel the mysteries of the cosmos.

12.3 Database Management

Unveiling the cosmic data realm with clarity:

12.3.1 SQL:

The universal language of databases, empowering astronomers to query and manage vast datasets effortlessly.

12.3.2 Normalization:

Organizing data orbits for efficiency, reducing redundancy and enhancing the integrity of astronomical databases.

12.3.3 Indexes:

Accelerating data retrieval at warp speed, ensuring astronomers swiftly access critical information.

12.3.4 Backup Strategies:

Guardians of celestial data, ensuring its safety through robust backup protocols.

12.3.5 Real-world Application:

Database management is the mission control of astrophysical data, ensuring a smooth voyage through the cosmos of information.

12.4 Data Mining Techniques

Embark on a cosmic data expedition:

12.4.1 Clustering:

Unveiling celestial patterns through grouping, revealing hidden structures in vast datasets.

12.4.2 Regression Analysis:

Predicting astronomical phenomena with precision, exploring the relationships between variables in the cosmos.

12.4.3   Decision Trees:

Navigating the cosmic terrain with binary choices, unraveling complex decision-making processes in astronomical data.

12.4.4   Neural Networks:

Mimicking the brain's power, unlocking the potential to recognize and interpret intricate patterns in the stars.

12.4.5   Association Rules:

Connecting the dots in cosmic correlations, uncovering meaningful associations between diverse astronomical entities.

12.4.6   Real-world Application:

Data mining is the telescope of the digital age, revealing insights and discoveries within the vast universe of information.

12.5   Data Access Protocols

Embark on a cosmic data journey with efficient protocols:

12.5.1   HTTP (Hypertext Transfer Protocol):

Streamlining data from telescopes to databases, like a cosmic web browser fetching information seamlessly.

12.5.2   FTP (File Transfer Protocol):

Transferring astronomical insights securely, ensuring data integrity over vast cosmic distances.

12.5.3   RESTful (Representational State Transfer):

Navigating the universe of data through standardized conventions, creating a harmonious interaction between celestial servers and clients.

12.5.4 GraphQL (Graph Query Language):

Unleashing the power of celestial queries, allowing astronomers to precisely request the data they need.

12.5.5 Real-world Application:

Data access protocols are the interstellar highways, facilitating the smooth flow of information across the astronomical data landscape.

12.6 Data Sharing Policies

Navigating the cosmos of data ethics:

12.6.1 Open Access:

Imagine a universe where knowledge flows freely. Astrophysical data, like starlight, should be accessible to all curious minds.

12.6.2 Authorship and Credit:

In the cosmic data realm, acknowledging contributors is crucial. Like constellations in the night sky, every author's role shines distinctly.

12.6.3 Curation Standards:

Maintaining the purity of the celestial data pool. Just as astronomers meticulously calibrate instruments, data curators ensure accuracy and reliability.

12.6.4 Data Embargoes:

Sometimes, like a cosmic mystery awaiting revelation, data might have a specific release date, adding a touch of suspense to astronomical discoveries.

12.6.5   Real-world Application:

Data sharing policies are the gravitational forces holding together the collaborative cosmos of astrophysical research.

12.7   Future Developments

Glimpsing into the cosmic horizon of data evolution:

12.7.1   Quantum Data Entanglement:

Picture data interconnected across vast cosmic distances, influenced by each other's state. Quantum-inspired data frameworks promise unprecedented correlations.

12.7.2   Neural Network Constellations:

Imagine artificial neural networks mimicking the complexity of galactic structures. Machine learning algorithms, like celestial bodies, evolving to process astronomical datasets.

12.7.3   Blockchain Astrophysics:

Securing the cosmos of data with the immutable ledger. Blockchain technology ensures transparency and authenticity in astrophysical data transactions.

12.7.4   Interstellar Data Transfer:

Efficiently transmitting colossal datasets across the cosmic void. Novel protocols resembling cosmic messengers to facilitate swift and secure interstellar data exchange.

12.7.5   Real-world Application:

The future of astrophysical data is a dynamic nebula, evolving with technological advancements, unlocking new realms of understanding the universe.

Chapter 13

Astrophysics and Society

13.1 Public Outreach

Bringing the cosmos to the public doorstep:

13.1.1 Starry Nights:

Engage communities with telescope events. Use the power of optics to unveil the beauty of celestial bodies, fostering a sense of wonder.

13.1.2 Astroedu-tainment:

Merge education and entertainment. Develop interactive programs, blending astronomy lessons with captivating shows and demonstrations.

13.1.3 Social Media Constellations:

Harness the gravitational pull of social platforms. Share bite-sized cosmic facts, captivating visuals, and live celestial events to create a celestial buzz.

13.1.4 Celestial Citizen Science:

Transform enthusiasts into contributors. Citizen science projects allow the public to participate in real astronomical research, turning stargazers into scientists.

13.1.5   Real-world Impact:

As constellations dot the night sky, public outreach constellations connect society with the vast wonders of the universe, fostering a cosmic perspective.

13.2   Science Education

Fueling curiosity, igniting minds:

13.2.1   Cosmic Classroom:

"From Stars to Students" - Integrate astronomy into school curricula, making the cosmos a dynamic part of science education.

13.2.2   Hands-On Universe:

Equip students with telescopes and DIY projects. Learn astronomy by doing, unleashing the power of tactile learning.

13.2.3   Astro Mentors:

Pair seasoned astronomers with students. Create mentorship programs, providing guidance and inspiration for budding astrophysicists.

13.2.4   Stellar Labs:

Transform classrooms into laboratories. Conduct experiments simulating cosmic phenomena, bridging theory with hands-on exploration.

13.2.5   Mathematical Orbits:

Decode the universe through equations. Integrate mathematical formulas seamlessly, making astrophysics a playground for mathematical minds.

13.2.6 Astrochemistry Corner:

Infuse chemistry with cosmic elements. Explore molecular compositions in space, connecting the dots between chemistry and astronomy.

13.2.7 Educational Nebula:

Where knowledge expands like a nebula, science education in astrophysics illuminates minds, creating a galaxy of informed thinkers.

13.3 Ethical Considerations

Navigating the Cosmos responsibly:

13.3.1 Stellar Ethics:

In the vast expanse of discovery, uphold ethical standards as guiding stars for astrophysicists and researchers.

13.3.2 Interstellar Collaboration:

Forge international collaborations, fostering unity in the pursuit of knowledge, transcending boundaries for the betterment of humanity.

13.3.3 Data Constellations:

Handle astronomical data with care. Develop protocols ensuring privacy and security, preventing unintended consequences.

13.3.4 Celestial Preservation:

Advocate for the protection of celestial environments. Consider the impact of space exploration on extraterrestrial ecosystems.

13.3.5 Inclusive Universe:

Ensure inclusivity in research and dissemination. Build bridges to make the benefits of astrophysics accessible to diverse communities.

13.3.6 Cosmic Diplomacy:

In the spirit of cosmic exploration, engage in diplomatic efforts, using astrophysics as a common ground for international relations.

13.3.7 Quantum Accountability:

Embrace accountability in astrophysical endeavors. Balance the quest for knowledge with a commitment to ethical conduct.

13.4 Policy Implications

Navigating the Cosmos within Policy Frameworks:

13.4.1 Celestial Governance:

Develop robust policies for the governance of celestial activities. Establish frameworks that balance exploration with environmental and ethical considerations.

13.4.2 Universal Collaboration Treaty:

Propose a treaty for international collaboration in space exploration. Define shared principles, responsibilities, and benefits for all participating nations.

13.4.3 Space Debris Mitigation Protocol:

Craft policies to address the growing issue of space debris. Implement strategies for responsible satellite deployment and orbital maintenance.

13.4.4   Astro-Ethical Guidelines:

Formulate ethical guidelines specific to astrophysical research. Address potential consequences, ensuring ethical conduct in the pursuit of knowledge.

13.4.5   Interplanetary Resource Management:

Create policies for the sustainable use of extraterrestrial resources. Balance scientific exploration with responsible resource utilization.

13.4.6   Cosmic Education Policy:

Advocate for policies that promote astrophysical education. Ensure widespread understanding of the significance of space exploration for society.

13.4.7   Astro-Innovation Incentives:

Introduce incentives for technological innovations in astrophysics. Encourage advancements that benefit both space exploration and terrestrial applications.

13.5   Technological Spin-Offs

Harnessing Astrophysical Innovation for Earthly Advancements:

13.5.1   Space Technologies on Earth:

Explore how technologies developed for space exploration benefit life on Earth. Examples include satellite imaging for weather forecasting and GPS for navigation.

13.5.2   Medical Applications:

Highlight medical advancements inspired by astrophysics. Discuss technologies like imaging devices originally designed for space missions now used in medical diagnostics.

13.5.3   Materials Science Breakthroughs:

Examine materials developed for space missions with applications on Earth. For instance, lightweight and durable materials finding use in everyday products.

13.5.4   Communication Innovations:

Explore how advancements in communication technology driven by space missions contribute to global connectivity and telecommunications.

13.5.5   Environmental Monitoring:

Discuss space-based tools for environmental monitoring. Showcase how satellite data aids in tracking climate change, deforestation, and natural disasters.

13.5.6   Energy Solutions:

Examine space-based solar power and its potential as a clean energy solution. Discuss the transfer of space-derived energy technologies to terrestrial applications.

13.5.7   Educational Outreach Tools:

Highlight technologies developed for space education programs. Discuss interactive tools and applications that make astrophysics more accessible to the public.

13.6   Collaborations and International Partnerships

Strengthening Global Bonds for Cosmic Exploration:

13.6.1   International Space Missions:

Explore how countries collaborate on space missions. Highlight successful joint ventures and the pooling of resources for shared scientific goals.

13.6.2 Space Agencies Collaboration:

Examine partnerships between major space agencies like NASA, ESA, ROSCOSMOS, etc. Discuss shared projects, technology exchange, and collaborative research.

13.6.3 Data Sharing Initiatives:

Highlight the importance of sharing astronomical data globally. Discuss international databases and platforms facilitating data exchange for the benefit of the global scientific community.

13.6.4 Telescope Collaborations:

Explore instances where countries jointly operate telescopes. Discuss how shared observatories lead to more comprehensive and diverse observations.

13.6.5 Space Exploration Technologies:

Examine international cooperation in developing advanced space technologies. Discuss collaborative efforts in spacecraft design, propulsion, and exploration systems.

13.6.6 Scientific Exchange Programs:

Highlight initiatives promoting the exchange of scientists and researchers across borders. Discuss how international collaboration enhances the diversity of expertise.

13.6.7 Educational Partnerships:

Explore collaborations in astrophysics education. Discuss joint programs, workshops, and initiatives fostering knowledge exchange among students globally.

13.6.8 Diplomatic Impacts:

Examine how astrophysical collaborations contribute to diplomatic relations. Highlight instances where space cooperation strengthens international ties.

13.7   Future Challenges

Navigating the Cosmic Frontier: Challenges and Solutions

13.7.1   Space Debris Mitigation:

Addressing the growing issue of space debris to ensure the sustainability of future space missions. Utilize mathematical models to calculate potential collision risks.

13.7.2   Sustainable Space Exploration:

Explore ways to make space exploration environmentally sustainable. Use equations to assess the impact of space missions on Earth and develop eco-friendly practices.

13.7.3   International Collaboration:

Enhancing global cooperation to tackle upcoming challenges. Discuss the mathematical frameworks for coordinating international efforts in addressing cosmic phenomena.

13.7.4   Space Policy and Governance:

Analyzing the mathematical aspects of formulating effective space policies. Discuss the equations governing responsible space exploration and resource utilization.

13.7.5   Public Engagement Strategies:

Developing mathematical models to maximize public engagement in astrophysics. Explore formulas to measure the impact of outreach programs and science communication.

13.7.6   Technological Advancements:

Anticipating mathematical challenges in adopting cutting-edge technologies. Discuss equations related to the development of faster propulsion systems, advanced telescopes, and data processing methods.

13.7.7 Education Accessibility:

Utilizing mathematical models to improve access to astrophysics education. Discuss formulas for designing inclusive educational programs and reducing barriers to entry.

13.7.8 Ethical Considerations:

Incorporate ethical dimensions into mathematical frameworks. Discuss equations that account for ethical considerations in space exploration and research.

Chapter 14

Frontiers in Astrophysics

14.1 Gravitational Waves

Ripples in Spacetime: Unveiling the Secrets of the Cosmos

14.1.1 Overview:

Gravitational waves are disturbances in the fabric of spacetime, propagating at the speed of light. They are generated by accelerating masses and carry information about cosmic phenomena.

14.1.2 Mathematical Foundation:

Einstein's Field Equations ($G_{\mu\nu} = 8\pi G T_{\mu\nu}$) describe the generation of gravitational waves. The wave equation ($\Box h_{\mu\nu} = \frac{16\pi G}{c^4} T_{\mu\nu}$) models their propagation.

14.1.3 Detection Techniques:

Utilize interferometry, where the change in distance between test masses is measured, to detect minuscule spacetime distortions caused by passing gravitational waves.

14.1.4 LIGO Experiment:

The Laser Interferometer Gravitational-Wave Observatory (LIGO) employs Michelson interferometers to detect gravitational waves. Equations govern the interpretation of interferometer data.

14.1.5 Binary Black Hole Mergers:

Mathematically describe the inspiral, merger, and ringdown phases of binary black hole systems using the post-Newtonian approximation and numerical relativity.

14.1.6 Einstein's Tensor:

Gravitational waves are represented by the Einstein tensor $(G_{\mu\nu})$. Explore the mathematical intricacies of this tensor and its role in describing spacetime curvature.

14.1.7 Astrophysical Significance:

Gravitational wave astronomy enables the study of black holes, neutron stars, and cosmological events. Discuss the equations used to extract information from observed signals.

14.1.8 Future Prospects:

Explore equations related to next-generation detectors like the Laser Interferometer Space Antenna (LISA) and their potential to unlock new realms of astrophysical knowledge.

14.2 Multimessenger Astronomy

Unveiling Cosmic Mysteries Through Multiple Messengers

14.2.1 Overview:

Multimessenger astronomy combines data from different cosmic messengers (e.g., electromagnetic waves, gravitational waves, neutrinos) to provide a comprehensive view of astrophysical phenomena.

14.2.2 Mathematical Foundation:

The combined analysis involves statistical methods (p-values, confidence intervals) and mathematical frameworks (likelihood functions) to infer the properties of astronomical sources.

14.2.3 Electromagnetic Waves:

Understand the equations governing the behavior of electromagnetic waves, including Maxwell's equations ($\nabla \cdot \mathbf{E} = \frac{\rho}{\varepsilon_0}$ and $\nabla \cdot \mathbf{B} = 0$).

14.2.4 Gravitational Waves:

Explore the interaction of spacetime with massive objects, leading to the generation of gravitational waves. Einstein's field equations play a crucial role.

14.2.5 Neutrinos:

Understand the fundamental properties of neutrinos and their detection mechanisms. Equations describe neutrino oscillations ($\nu_\alpha = \sum_i U_{\alpha i} \nu_i$).

14.2.6 Joint Observations:

Mathematically model joint observations, incorporating data from various messengers. Bayesian statistics aids in combining information from different sources.

14.2.7 Advanced LIGO and Virgo:

Discuss upgrades to gravitational wave detectors like Advanced LIGO and Virgo. Equations describe the improved sensitivity and capabilities.

14.2.8 Astrophysical Applications:

Explore equations related to the study of compact binary mergers, supernovae, and other astrophysical events using multimessenger data.

14.2.9 Cosmic Ray Detection:

Understand the equations behind the detection of cosmic rays, high-energy particles originating from various astrophysical sources.

14.2.10   Future Directions:

Examine mathematical frameworks for upcoming multimessenger observatories, emphasizing the synergy between different observational techniques.

14.3   Astroinformatics

Navigating the Cosmos with Data and Algorithms

14.3.1   Overview:

Astroinformatics harnesses computational techniques to analyze vast datasets, enabling the extraction of meaningful insights from the cosmos.

14.3.2   Mathematical Foundation:

Key mathematical tools include algorithms for data mining, machine learning models (e.g., neural networks), and statistical methods for robust analysis.

14.3.3   Data Representation:

Understand the mathematical representation of astronomical data, employing matrices, tensors, and mathematical structures suitable for algorithmic processing.

14.3.4   Machine Learning:

Explore the equations behind machine learning algorithms, such as the training process of neural networks ($W_{ij} \leftarrow W_{ij} - \alpha \frac{\partial L}{\partial W_{ij}}$).

14.3.5   Statistical Analysis:

Utilize statistical equations for hypothesis testing (t-tests, ANOVA) and regression analysis to quantify uncertainties and relationships within astronomical datasets.

14.3.6 Computational Simulations:

Mathematical modeling equations describe the simulation of astrophysical phenomena, enabling predictions and comparisons with observational data.

14.3.7 Data Fusion:

Explore equations for integrating diverse datasets, merging observations from telescopes, satellites, and other instruments for a comprehensive view.

14.3.8 Big Data Challenges:

Address mathematical challenges in handling large astronomical datasets, including scalable algorithms and parallel processing techniques.

14.3.9 Visualization Techniques:

Equations related to data visualization, employing graphical representations, color maps, and other mathematical methods to convey complex information.

14.3.10 Astroinformatics Platforms:

Discuss the mathematical underpinnings of astroinformatics platforms, emphasizing their role in processing, analyzing, and sharing astronomical data.

14.3.11 Future Prospects:

Explore emerging mathematical methodologies and equations driving the future of astroinformatics, including real-time data processing and advanced algorithms.

14.4 Artificial Intelligence in Astrophysics

Unleashing AI for Cosmic Discoveries

14.4.1   Overview:

Artificial Intelligence (AI) revolutionizes astrophysics by automating tasks, analyzing data, and unveiling hidden patterns in the vastness of the universe.

14.4.2   Mathematical Foundation:

Central equations include those governing neural networks, reinforcement learning algorithms, and optimization functions employed in training AI models.

14.4.3   Data Processing:

Mathematical equations illustrate preprocessing steps ($X_{\text{normalized}} = \frac{X-\mu}{\sigma}$) and feature extraction techniques crucial for AI-based analysis of astronomical data.

14.4.4   Neural Networks:

Explore the equations behind neural networks, encompassing activation functions ($a = \sigma(z) = \frac{1}{1+e^{-z}}$) and backpropagation for learning.

14.4.5   Machine Learning Applications:

Equations exemplify supervised and unsupervised learning algorithms applied to astrophysical problems, such as classification and clustering.

14.4.6   Predictive Modeling:

Mathematical expressions underpin predictive modeling, facilitating the forecasting of celestial events or behaviors using AI-driven algorithms.

14.4.7   Optimization:

Understand optimization equations ($W_{\text{new}} = W_{\text{old}} - \alpha \frac{\partial J}{\partial W}$) for refining AI models to enhance accuracy and efficiency.

14.4.8 Transfer Learning:

Explore mathematical concepts underpinning transfer learning, enabling AI models to leverage knowledge gained in one astrophysical domain for another.

14.4.9 Interpretability:

Equations illustrate methods for making AI decisions interpretable, ensuring transparency and understanding of the insights derived.

14.4.10 AI in Survey Analysis:

Mathematical foundations of using AI to analyze vast survey datasets, speeding up the identification of celestial objects and phenomena.

14.4.11 Ethical Considerations:

Discuss equations that may guide ethical considerations in AI applications, emphasizing fairness, accountability, and transparency.

14.4.12 Future Horizons:

Explore emerging mathematical frameworks for AI advancements in astrophysics, anticipating breakthroughs in autonomous exploration and discovery.

14.5 Quantum Astrophysics

Navigating Cosmic Realms with Quantum Insights

14.5.1 Overview:

Quantum astrophysics fuses the principles of quantum mechanics with the cosmic expanse, unlocking new dimensions in our understanding of the universe.

14.5.2   Mathematical Foundation:

Key equations include Schrödinger's equation ($i\hbar\frac{\partial}{\partial t}\Psi = \hat{H}\Psi$) applied to celestial systems, emphasizing wave functions and quantum states.

14.5.3   Wave-Particle Duality:

Explore the duality concept ($p = \frac{h}{\lambda}$), illustrating how particles in space exhibit both wave and particle characteristics.

14.5.4   Quantum Entanglement:

Equations depict entanglement phenomena, highlighting the interconnected nature of particles across vast cosmic distances.

14.5.5   Quantum Tunneling in Stars:

Mathematical expressions elucidate quantum tunneling ($T \approx e^{-2Gm(R-r)/\hbar c}$), a phenomenon influencing nuclear fusion within stars.

14.5.6   Quantum Computing in Astrophysics:

Understand quantum gates and algorithms ($|\psi\rangle = U|\phi\rangle$), showcasing the potential of quantum computing for complex astrophysical simulations.

14.5.7   Quantum Gravity Theories:

Explore equations from quantum gravity theories, like loop quantum gravity or string theory, offering insights into the fundamental nature of space and time.

14.5.8   Quantum Black Holes:

Mathematical foundations reveal quantum aspects of black holes, addressing information paradoxes and Hawking radiation ($T = \frac{\hbar c^3}{8\pi GMk}$).

14.5.9   Cosmic Microwave Background in Quantum Terms:

Express the cosmic microwave background using quantum principles, emphasizing its significance in understanding the early universe.

14.5.10   Quantum Astrophysics Experiments:

Illustrate equations underpinning experiments probing quantum aspects of astrophysical phenomena, bridging the macro and quantum worlds.

14.5.11   Quantum Uncertainty in Cosmic Measurements:

Discuss Heisenberg's uncertainty principle ($\Delta x \Delta p \geq \frac{\hbar}{2}$) and its implications for precise cosmic measurements.

14.5.12   Future Frontiers:

Explore emerging mathematical frameworks in quantum astrophysics, anticipating discoveries that may reshape our cosmic comprehension.

14.6   Astrochemistry

Unveiling Cosmic Chemical Mysteries

14.6.1   Overview:

Astrochemistry delves into the chemical processes shaping celestial bodies, unraveling the cosmic chemistry that governs our universe.

14.6.2   Mathematical Foundation:

Key equations involve chemical kinetics ($\frac{d[A]}{dt} = -k[A]$), quantum chemistry principles, and thermodynamics applied to astronomical environments.

14.6.3   Molecular Formation in Space:

Illustrate reactions ($2H_2 + O_2 \rightarrow 2H_2O$) leading to the formation of complex molecules in interstellar space, fostering the birth of stars and planets.

14.6.4   Stellar Nucleosynthesis:

Equations describe nuclear reactions within stars ($4^1H \rightarrow^4 He + 2e^+ + 2\nu_e$), outlining the creation of elements through fusion processes.

14.6.5   Chemical Composition of Exoplanetary Atmospheres:

Quantify chemical equilibrium ($\frac{[C]}{[CO]}$ = constant) to understand the atmospheres of exoplanets and their potential habitability.

14.6.6   Isotopic Ratios as Cosmic Chronometers:

Explore isotopic ratios ($\frac{^{87}Sr}{^{86}Sr}$) as cosmic clocks, helping determine the age of celestial bodies and tracing their evolutionary history.

14.6.7   Chemical Signatures in Cosmic Microwave Background:

Connect chemical imprints to cosmic microwave background, revealing insights into the early universe's elemental composition.

14.6.8   Molecular Cloud Dynamics:

Describe cloud collapse processes ($\frac{d\rho}{dt} \propto -\rho^{1/2}$) leading to star formation, emphasizing the role of chemistry in shaping stellar systems.

14.6.9   Complex Organic Molecules in Space:

Highlight complex reactions ($HC_3N + C_2H_2 \rightarrow C_5H_2N + H$) producing organic compounds in the vastness of interstellar space.

14.6.10   Chemical Fingerprints of Cosmic Events:

Examine chemical signatures ($^{56}Ni \rightarrow ^{56}Co + e^+ + \nu_e$) from cosmic events like supernovae, aiding in the understanding of stellar processes.

14.6.11   Future Frontiers:

Anticipate advancements in astrochemical models and experiments, paving the way for a deeper comprehension of the chemical intricacies shaping our cosmos.

14.7   Theoretical Astrophysics

Unveiling the Universe through Mathematical Insights

14.7.1   Overview:

Theoretical astrophysics employs mathematical models to decipher cosmic phenomena, unraveling the fundamental principles governing our universe.

14.7.2   Mathematical Foundation:

Theoretical astrophysics relies on a plethora of equations, including Newton's law of gravitation ($F = G\frac{m_1 m_2}{r^2}$), Einstein's field equations, and Maxwell's equations for electromagnetic fields.

14.7.3   Cosmological Models:

Utilize Friedmann equations ($H^2 = \frac{8\pi G}{3}\rho - \frac{k}{a^2}$) to predict the large-scale structure and evolution of the universe, exploring scenarios like inflation.

14.7.4   Black Hole Dynamics:

Describe the geometry around black holes using the Schwarzschild metric ($ds^2 = -\left(1 - \frac{2GM}{c^2 r}\right)c^2 dt^2 + \frac{1}{1-\frac{2GM}{c^2 r}}dr^2 + r^2(d\theta^2 + \sin^2\theta d\phi^2)$).

14.7.5   Gravitational Wave Formulation:

Highlight the mathematics of gravitational wave generation, propagation, and detection, encapsulated in the quadrupole formula.

14.7.6   Quantum Astrophysics:

Integrate quantum mechanics with astrophysical phenomena, employing equations like the Schrödinger equation to understand the behavior of matter on cosmic scales.

14.7.7   Stellar Structure Equations:

Explore the polytropic equation ($\frac{dP}{dr} = -\frac{GM\rho}{r^2}$), providing insights into the internal structure and evolution of stars.

14.7.8   Galactic Dynamics:

Apply the Jeans equations to model the distribution of matter in galaxies, unraveling the gravitational interactions shaping their intricate structures.

14.7.9   Time-Dependent Stellar Evolution:

Investigate the set of coupled differential equations governing stellar evolution, predicting the life cycles of stars and their eventual fates.

14.7.10   Dark Energy and Modified Gravity Theories:

Introduce modified theories of gravity ($f(R)$ gravity) to explain cosmic acceleration, unveiling alternatives to dark energy.

14.7.11   Theoretical Insights into Multiverse Theories:

Delve into the mathematical frameworks of multiverse theories, exploring the possibility of diverse and interconnected universes.

14.7.12 Future Frontiers:

Anticipate advancements in theoretical models, expecting breakthroughs that will deepen our understanding of the cosmos and challenge existing paradigms.

Chapter 15

Conclusion

15.1 Summary

Unveiling the Cosmic Tapestry: A Recap

15.1.1 Cosmology:

From the expansion of the universe ($H = \frac{\dot{a}}{a}$) to the enigmatic dark energy ($\rho_{\mathrm{DE}} = $ constant), cosmology illuminates the grand narrative of our cosmic evolution.

15.1.2 High-Energy Astrophysics:

Navigating black holes ($R_{\mu\nu} - \frac{1}{2}g_{\mu\nu}R + g_{\mu\nu}\Lambda = \frac{8\pi G}{c^4}T_{\mu\nu}$), neutron stars, and gamma-ray bursts, high-energy astrophysics unfolds the energetic ballet of celestial bodies.

15.1.3 Observational Techniques:

Telescopes, spectroscopy ($E = h\nu$), and radio astronomy unveil the cosmos, providing snapshots of distant galaxies and analyzing their chemical compositions.

15.1.4 Exoplanets and Astrobiology:

From Drake's equation to exoplanetary atmospheres, the quest for habitable worlds and extraterrestrial life pushes the boundaries of astrobiology.

15.1.5   High-Performance Computing:

Numerical methods, parallel computing, and data visualization propel astrophysics into the computational era, simulating the cosmos with unprecedented accuracy.

15.1.6   Astrophysical Instrumentation:

From radio telescopes to adaptive optics, cutting-edge instruments capture cosmic wonders, enabling us to witness the celestial ballet with unparalleled clarity.

15.1.7   Astrophysical Constants and Units:

Grasping the language of the cosmos, astronomical constants and unit systems provide the fundamental framework for precise astronomical calculations.

15.1.8   Astrophysical Data and Databases:

Mining vast datasets, astronomers access repositories, employ data mining techniques, and adhere to robust data sharing policies to propel discoveries.

15.1.9   Astrophysics and Society:

Engaging the public, fostering science education, and addressing ethical considerations, astrophysics intertwines with society, leaving an indelible mark on policy and technology.

15.1.10   Frontiers in Astrophysics:

From gravitational waves and multimessenger astronomy to artificial intelligence and quantum astrophysics, the frontiers beckon, promising new realms of understanding.

15.1.11   Conclusion:

As we stand at the precipice of discovery, the cosmic odyssey continues. Astrophysics, both a science and a cultural endeavor, unveils the marvels of the universe, inspiring generations to explore the unknown.

15.2 Challenges Ahead

15.2.1 Navigating Dark Horizons:

The cosmic voyage faces challenges as vast as the universe itself. From understanding the nature of dark matter ($F = ma$) to deciphering the cosmic microwave background radiation, uncharted territories beckon.

15.2.2 Quantum Quandaries:

In the subatomic realm ($\Delta x \Delta p \geq \frac{\hbar}{2}$), the interplay of quantum mechanics and gravity poses profound questions. Bridging these theories stands as a monumental challenge for the next generation of physicists.

15.2.3 Cosmic Acceleration:

Decoding the acceleration of the universe requires unraveling the mysteries of dark energy ($pV = nRT$). New models and innovative approaches must be forged to grapple with the elusive force propelling cosmic expansion.

15.2.4 Technological Frontiers:

As technology propels us forward, embracing challenges in data handling, quantum computing, and AI becomes imperative ($E = mc^2$). These tools are the beacons guiding us through the immense sea of astrophysical information.

15.2.5 Sustainability in Space Exploration:

The quest for knowledge extends beyond Earth, demanding sustainable practices in space exploration. From rocket propellants to life support systems, astrophysical endeavors must echo the harmony of celestial bodies.

15.2.6 Global Collaboration:

Challenges are met with collective endeavors. Emphasizing international partnerships ($F_{\text{net}} = ma$) ensures that humanity's pursuit of cosmic understanding transcends borders, fostering a shared cosmic legacy.

15.2.7 Educational Empowerment:

Equipping the next generation with the tools to explore the cosmos ($W = Fd$) is crucial. Science education and outreach initiatives will mold the astronomers, engineers, and dreamers who will lead the cosmic charge.

15.2.8 Innovation in Astrobiology:

The search for extraterrestrial life ($P_{\text{ideal}} = nRT$) faces challenges that demand innovative approaches. Astrobiology's evolution requires interdisciplinary collaboration and cutting-edge methodologies.

15.2.9 Preserving Dark Skies:

Protecting the sanctity of our observational platforms, preserving dark skies ($\lambda = \frac{c}{f}$), is a shared responsibility. Mitigating light pollution ensures an unobstructed view into the depths of the universe.

15.2.10 Ethical Dimensions:

As we unravel the fabric of the cosmos, ethical considerations ($\Sigma F = ma$) must guide our journey. Responsible science, data sharing, and technology deployment will shape the ethical contours of astrophysical exploration.

15.2.11 Closing Cosmic Chapters:

In facing these challenges, the cosmic odyssey takes on new dimensions. The chapters ahead promise both the thrill of discovery and the satisfaction of overcoming the hurdles that propel us toward a deeper understanding of the universe.

15.3 Inspiration for Future Research

15.3.1 Cosmic Puzzles:

As we glimpse into the vastness of the universe, unresolved puzzles beckon future explorers ($E = mc^2$). From the nature of dark matter to the enigma of cosmic inflation, these mysteries stir the curiosity of the cosmic detective.

15.3.2 Quantum Cosmos:

The interplay of quantum mechanics and cosmology unfolds a cosmic theater ($\Psi(x,t)$). Exploring quantum aspects on cosmic scales and deciphering the quantum origins of the universe present enticing avenues for future research.

15.3.3 Time Travel and Wormholes:

Manipulating spacetime ($G_{\mu\nu} + \Lambda g_{\mu\nu} = \frac{8\pi G}{c^4} T_{\mu\nu}$) invites investigation into the feasibility of time travel and the existence of traversable wormholes. These temporal frontiers pose theoretical challenges and practical possibilities.

15.3.4 Exoplanet Biosignatures:

The search for life beyond Earth ($CH_4 + 2O_2 \rightarrow CO_2 + 2H_2O$) hinges on understanding exoplanet atmospheres. Detecting biosignatures and characterizing habitable zones offer a new realm for astrobiological exploration.

15.3.5 AI-Driven Astrophysics:

Embracing artificial intelligence ($NeuralNetwork \rightarrow$ Discovery) transforms data analysis and pattern recognition. The integration of AI promises breakthroughs in identifying celestial objects, classifying galaxies, and predicting cosmic phenomena.

15.3.6 Harnessing Dark Energy:

Unraveling the secrets of dark energy ($pV = nRT$) opens the door to innovative energy solutions. Understanding this cosmic force may inspire advancements in harnessing dark energy for sustainable technologies.

15.3.7 Quantum Astrophotography:

Applying quantum principles ($E = hf$) to astrophotography could revolutionize image capture. Quantum sensors and entangled photon pairs may offer unprecedented precision in capturing distant cosmic phenomena.

15.3.8   Gravitational Wave Communication:

Exploring the potential of gravitational waves ($GW \rightarrow$ Data Transmission) for communication may redefine long-distance communication in space. Gravitational wave signals could enable new methods of interstellar information exchange.

15.3.9   Synthetic Astrophysics:

Simulating astrophysical scenarios ($Numerical Simulation \rightarrow$ Virtual Universe) in unprecedented detail allows researchers to conduct experiments in a synthetic cosmos. Virtual observatories and simulated universes offer controlled environments for testing hypotheses.

15.3.10   Galactic Internet:

As we extend our reach into the cosmos, envisioning a galactic internet ($Quantum Entanglement \rightarrow$ Instantaneous Communication) challenges us to rethink the fabric of cosmic connectivity. Quantum entanglement might hold the key to instantaneous communication across vast cosmic distances.

15.3.11   Dark Skies Preservation:

Preserving dark skies ($\lambda = \frac{c}{f}$) becomes a critical mission. Advocating for regulations and technologies that minimize light pollution ensures the continued clarity of our view into the celestial wonders.

15.3.12   Interdisciplinary Collaboration:

The future of astrophysics lies in collaboration ($\Sigma F = ma$). Crossing disciplinary boundaries, merging physics with biology, computer science, and engineering, promises a holistic understanding of the cosmos.

15.3.13   Energizing the Cosmic Curiosity:

Inspiration for future research resides in energizing the cosmic curiosity of the next generation. Educational initiatives, outreach programs, and fostering a sense of wonder will drive future scientists and explorers to push the boundaries of knowledge.

15.4 Closing Thoughts

15.4.1 Celestial Harmony:

As we bid adieu to the cosmic symphony ($\omega = \frac{2\pi}{T}$), let's appreciate the harmony of celestial bodies. Each planet, star, and galaxy contributes a unique note, creating the cosmic melody that resonates through the vastness of space.

15.4.2 Time's Dance:

In the grand ballroom of the universe, time waltzes ($t = \frac{d}{v}$) gracefully. From the elegant orbits of planets to the explosive ballet of supernovae, every cosmic event is a dance choreographed by the laws of physics.

15.4.3 Stellar Legacy:

Stars, the celestial poets ($H - f = \frac{\Delta E}{\lambda}$), script their verses across the cosmic canvas. Their radiance paints stories of birth, evolution, and eventual farewell, leaving behind elements that will shape new cosmic narratives.

15.4.4 Galactic Tapestry:

Our view of the cosmos is a tapestry ($\sigma = \frac{m}{A}$) woven with the threads of galaxies. Each spiral arm, elliptical swirl, and irregular smudge contributes to the rich visual texture of the cosmic masterpiece.

15.4.5 Eternal Cosmic Flux:

In the river of spacetime, everything flows ($\nabla \cdot \mathbf{E} = \frac{\rho}{\varepsilon_0}$). From the birth of galaxies to the merging of black holes, the cosmic flux eternally shapes the landscape of the universe.

15.4.6 Quantum Whispers:

In the quantum realm, particles whisper secrets ($\Psi(x,t)$) that guide the cosmic narrative. Quantum fluctuations, entanglement, and the probabilistic dance of particles add an intriguing layer to the cosmic story.

15.4.7   Cosmic Alchemy:

Celestial laboratories ($n_1\lambda_1 = n_2\lambda_2$) perform alchemy, transforming light into spectral fingerprints. Analyzing these signatures unveils the cosmic chemistry that forges the building blocks of the universe.

15.4.8   Interstellar Resilience:

Amidst the cosmic storms ($F = ma$), galaxies and star clusters exhibit resilience. The delicate balance of gravitational forces and kinetic energies demonstrates the interstellar tenacity that withstands the cosmic tempest.

15.4.9   Universal Gratitude:

In the vast cosmic expanse, let's express gratitude ($\int_0^\infty e^{-x^2} dx = \frac{\sqrt{\pi}}{2}$). Grateful for the photons that traveled eons to paint the night sky and for the cosmic phenomena that reveal the wonders of the universe.

15.4.10   Beyond the Horizon:

As we contemplate the cosmic horizon, let's remember ($E = mc^2$) that our journey in astrophysics is a perpetual exploration. Beyond the observable, the universe unfolds mysteries that beckon future generations to venture into the cosmic unknown.